Use of AI, Robotics, and Modern Tools to Fight Covid-19

RIVER PUBLISHERS SERIES IN AUTOMATION, CONTROL AND ROBOTICS

Series Editors

ISHWAR K. SETHI
Oakland University
USA

TAREK SOBH
University of Bridgeport
USA

FENG QIAO
Shenyang JianZhu University
China

Indexing: all books published in this series are submitted to the Web of Science Book Citation Index (BkCI), to SCOPUS, to CrossRef and to Google Scholar for evaluation and indexing.

The "River Publishers Series in Automation, Control and Robotics" is a series of comprehensive academic and professional books which focus on the theory and applications of automation, control and robotics. The series focuses on topics ranging from the theory and use of control systems, automation engineering, robotics and intelligent machines.

Books published in the series include research monographs, edited volumes, handbooks and textbooks. The books provide professionals, researchers, educators, and advanced students in the field with an invaluable insight into the latest research and developments.

Topics covered in the series include, but are by no means restricted to the following:

- Robots and Intelligent Machines
- Robotics
- Control Systems
- Control Theory
- Automation Engineering

For a list of other books in this series, visit www.riverpublishers.com

Use of AI, Robotics, and Modern Tools to Fight Covid-19

Editors

Arpit Jain

University of Petroleum and Energy Studies, India

Abhinav Sharma

University of Petroleum and Energy Studies, India

Jianwu Wang

University of Maryland, USA

Mangey Ram

Graphic Era Deemed to be University, India

River Publishers

Published, sold and distributed by:
River Publishers
Alsbjergvej 10
9260 Gistrup
Denmark

www.riverpublishers.com

ISBN: 9788770224437 (Hardback)
          9788770224420 (Ebook)

Contents

Preface

The COVID-19 has hit the world at a gigantic scale. Modern technological tools such as artificial intelligence (AI), robotics, and Internet of Things (IoT) have served the mankind at various stages in combating the fight against the pandemic. This book focuses on the application of these innovative tools and their role in the COVID-19 crisis fight. The organization of the book is as follows.

Chapter 1 presents a brief history of pandemics ranging from prehistoric pandemics to modern day pandemics along with the detailed status of COVID-19. This chapter also discusses the impact of pandemic in various aspects.

Chapter 2 is focuses on tracing the origin of COVID-19. The chapter presents brief history of SARS-CoV where its origin and variants have been discussed.

Chapter 3 discusses the role of modern technology in combating the fight against COVID-19. The chapter focuses on various developments across the globe related primarily to AI, which are deployed in various domains including prediction, diagnostics, vaccine development, contact tracing, and many more.

Chapter 4 discusses the development of technological innovations from India in combating the fight against COVID-19. The chapter focuses on multitude of technologies like: robotics, unmanned aerial vehicles (UAVs), AI, surveillance systems, and CFD (Computational Fluid Dynamics) techniques.

Chapter 5 provides a survey of robotic systems deployed in real time to combat the COVID-19 pandemic fight. The paper discusses the various aspects of robot assistance in healthcare and various challenges that are faced during a robotic system design.

Chapter 6 uses the reproduction number of the virus and susceptible exposed infectious recovered death cumulative structure to predict the number of COVID-19 cases in India. Authors took into account the phased lockdown and unlock strategies implemented in India. The authors have also predicted the reproduction number.

Chapter 7 utilizes the time-series distributions to predict the number of cases in Baltimore County and Prince George's County. The authors used time series analysis to analyze COVID-19 spread at various locations to study the overall patterns that emerge. Piecewise aggregate approximation (PAA), matrix profiles (MPs), and time series discretization methods were used. The analysis and predictions have a high impact on policy making.

Chapter 8 focuses on optimization of nurse scheduling at hospitals using evolutionary algorithms. The scheduling of frontline workers is extremely important to relieve the system from extra stress and for mental and physical wellbeing of the frontline workers. Performance of the grey wolf optimization and particle swarm optimization methods are compared and conclusions are drawn.

Chapter 9 focuses on the role of future smart city in pandemic control. Wireless sensor networks and Internet of Things are the backbone of a smart city. The authors have simulated the case of a smart city using CupCarbon™ tool. The authors traced the movement of infectious nodes and the spread has been analyzed.

Chapter 10 focuses on the effect on trade and economic impacts of the pandemic. The authors also analyzed the technical barriers to trade (TBT) and sanitary and phytosanitary (SPS) measures as directed by WTO.

Chapter 11 addresses the impact and interplay of the COVID-19 pandemic and climate change. The two disruptors have been compared via detailed analysis of key parameters such as CO_2 and greenhouse gas (GHG) emission levels. The dataset for major countries like USA, EU, China, and Russia have been used for comparison analysis. The authors have concluded the manuscript by recommending the best practices induced by the "new normal" which will be effective in tackling the climate change and sustainable future.

Chapter 12 assesses the impact of the pandemic in the field of education and academics in India. The pandemic timeline in India has been utilized in order to assess the impact on higher education and various virtual platforms have been discussed which can be utilized by the students for continuing education in the challenging times.

Chapter 13 focuses on the impact of pandemic on various industrial sectors and the role of virtual reality to tackle the situation. The authors also discuss the challenges in implementation of VR techniques and implications of further research.

This research book caters to a wide domain of researchers working in the domain of AI, robotics, and IoT/wireless sensor network (WSN). The COVID-19 pandemic has widely disrupted the human life and multidisciplinary research is more significant than ever. This book covers AI-based prediction models to robotic system applications with respect to the mankind combat to COVID-19. This book not only serves the academic and technical purposes but also serves the reader of general domains to gain some insights of the latest developments in the technological framework and its impact on healthcare. This book can also serve as a valuable reference for academics, mechanical, mechatronics, computer science, information technology and industrial engineers, environmental sciences, as well as researchers in related subjects.

Dehradun, India

Editors:
Arpit Jain
Abhinav Sharma
Jianwu Wang
Mangey Ram

Acknowledgements

The Editor acknowledges River Publishers for this opportunity and professional support. Our special thanks to Mr. Rajeev Prasad, Ms. Junko Nakajima River Publishers for the excellent support, provided us to complete this book.

Thanks to the chapter authors and reviewers for their valuable contributions to the book.

Arpit Jain
University of Petroleum and Energy Studies, India

Abhinav Sharma
University of Petroleum and Energy Studies, India

Jianwu Wang
University of Maryland, USA

Mangey Ram
Graphic Era Deemed to be University, India

List of Contributors

Abhinav Sharma, *Department of Electrical and Electronics Engineering, University of Petroleum and Energy Studies, Dehradun, Uttarakhand 248007, India*

Anamika Rana, *Maharaja Surajmal Institute of Technology, Janakpuri, Delhi 110058, India*

Anna M. Rybakova, *Moscow State Institute of International Relations University, Moscow, Russia*

Arpit Jain, *Department of Electrical and Electronics Engineering, University of Petroleum and Energy Studies, Dehradun, Uttarakhand 248007, India*

Aslesha Bodavula, *Kings Cornerstone International College, Chennai, Tamil Nadu, India*

Chhaya Kulkarni, *Department of Information Systems, University of Maryland, Baltimore County, Baltimore, MD, USA*

Hardik Patel, *Department of Information and Communication Technology, Pandit Deendayal Petroleum University, Gandhinagar, Gujarat 382007, India*

Iosif Z. Aronov, *Moscow State Institute of International Relations University, Moscow, Russia*

Vibhu Jately, *MCAST Energy Research Group, Malta College of Arts, Science and Technology, MCAST, Paola, Malta*

Rajendra Kumar Jatley, *Department of Electrical and Computer Engineering, Wollega University, Nekemte, Ethiopia*

Jyoti Joshi, *Department of Electrical Engineering, College of Technology, G. B. Pant University of Agriculture and Technology, Pantnagar, Uttarakhand 263153, India*

Kamal Rawat, *Meerut Institute of Engineering & Technology, Meerut 250005, India*

Kuldeep Panwar, *General Electric Power, Noida 201303, India*

Mangey Ram, *Department of Mathematics, Graphic Era University, Dehradun, Uttarakhand 248002, India*

Meera C. S., *Department of Electrical and Electronics Engineering, University of Petroleum and Energy Studies, Dehradun, Uttarakhand 248007, India*

Mohendra Roy, *Department of Information and Communication Technology, Pandit Deendayal Petroleum University, Gandhinagar, Gujarat 382007, India*

Nataliia M. Galkina, *International Trade and Integration (ITI) Research Center, Moscow, Russia*

Neeraj Bisht, *Department of Mechanical Engineering, G.B. Pant University of Agriculture Engineering, Pantnagar 263153, India*

Paawan Sharma, *Department of Information and Communication Technology, Pandit Deendayal Petroleum University, Gandhinagar, Gujarat 382007, India*

Pinisetti Swami Sairam, *School of Business, Woxsen University, Hyderabad, Telangana, India*

Prashant Kumar Dwivedi, *The Hi-tech Robotic Systemz Ltd, Gurgaon, Haryana 122001, India*

Puneet Joshi, *Department of Electrical Engineering, Rajkiya Engineering College, Ambedkar Nagar, Akbarpur, Uttar Pradesh 224122, India*

Rakhi Pandey, *Department of Humanities and Social Sciences, Jaypee University, Anoopshahr, Uttar Pradesh 203390, India*

Sandipan Dey, *Department of Information Systems, University of Maryland, Baltimore County, Baltimore, MD, USA*

Sirisha Velampalli, *CR Rao AIMSCS, University of Hyderabad, Gachibowli, Telangana 500046, India*

Son Vu Truong Dao, *International University, Vietnam National University, Ho Chi Minh City, Vietnam*

Sriperumbuduru Srilaya, *CR Rao AIMSCS, University of Hyderabad, Gachibowli, Telangana 500046, India*

Supriya Pandey, *Sr. Engineer, General Electric Power, Noida 201303, India*

Sushma Malik, *Institute of Innovation in Technology & Management, Janakpuri, Delhi 110058, India*

Swati Shukla, *Department of Mathematics, Gorakhpur University, Gorakhpur, Uttar Pradesh 273009, India*

Tan Nhat Pham, *International University, Vietnam National University, Ho Chi Minh City, Vietnam*

Vandana Janeja, *Department of Information Systems, University of Maryland, Baltimore County, Baltimore, MD, USA*

Vidushee Nautiyal, *Department of Education, Uttaranchal College of Education, Dehradun, Uttarakhand 248001, India*

List of Figures

List of Tables

List of Notations and Abbreviations

RNA   Ribonucleic acid
COVID-19   Corona virus
SARS-CoV   Severe acute respiratory syndrome coronavirus
NSP   Nurse scheduling problem
GWO   Grey wolf optimization
PSO   Particle swarm optimization

1

The History of Pandemics and Evolution So Far

Puneet Joshi[1] and Swati Shukla[2]

[1] Department of Electrical Engineering, Rajkiya Engineering College, Ambedkar Nagar, Akbarpur, Uttar Pradesh 224122, India.
[2] Department of Mathematics, Gorakhpur University, Gorakhpur, Uttar Pradesh 273009, India.
Corresponding Author: Puneet Joshi, drpuneetj@recabn.ac.in.

Abstract

In the realm of infectious diseases, a pandemic is the worst-case scenario. When an epidemic is spread beyond a country's borders, then it is officially called pandemic. Humankind has always seen communicable diseases, but shift to aggregarion life made epidemic and later pandemic more possible. Malaria, tuberculosis, leprosy, influenzae [1, 2], and smallpox were some of them. The more civilized humans became, the more likely the pandemics became – like the first cholera pandemic in 1817 originated in Russia. In recent history, the world's one of the deadliest pandemics in 1918 was Spanish flu which struck 50 million deaths in Europe, USA, and parts of Asia. In 1981 AIDS, and then the world witnessed SARS-CoV-1 in 2003 in China, a form of coronavirus, a name which helps us recognize this new one of the most contagious pandemics the world has ever witnessed, called SARS-CoV-2 or popularly called COVID-19 or, as some say, Wuhan virus. SARS-CoV-2 was reported by China in late December 2019 with pneumonia-like symptoms of unknown origin. Without a vaccine available, it spread beyond borders, to more than 163 countries. COVID-19 has revealed vulnerabilities in the global communities' response to outbreaks of viruses, which has damaged the world's economy to the worst recession since the Great Depression. Shift of power and bringing the world to a standstill is something which none has ever imagined. Deaths, broken families, helplessness, fear and anxiety, uncertainty, and many more emotions were heaped on mankind. Economies suffered but the planet breathed, humans stopped but wildlife played. Life under COVID-19 brought blue skies and clean air, and many suffered but then many held them. Humans sought to learn and strive, sufferings created fighters, roadblocks paved new paths, and this horrendous experience changed the outlook and paved the perspective for a new world after COVID-19.

1.1 Introduction

Ailments and illness have tormented humankind since the earliest days of our mortal flaw. Yet, moving to aggregarion, communities augmented the spread drastically. The more edified the people became with exotic trading avenues with augmented commerce with various populaces of individuals, faunae, and ecosystems, the more probable pandemics became. There have been numerous critical ailment outbursts and pandemics chronicled, including Spanish Flu, Hong Kong Flu, SARS, COVID-19, Ebola, Zika, etc. The expression "pandemic" has barely been characterized by numerous clinical writings; however, there are some significant attributes of a pandemic, including varied geographic reach, illness development, novelty, brutality, high mortality rates and instability, reduce populace immunity, and contagious and infectious, which assist to comprehend the problem better in the event that we analyze similitudes and contrasts amongst them. The pandemic-related causalities can be correlated to tremendous adverse effects on wellbeing, financial system, civilization, and safety of national and worldwide publics. Also, they have instigated noteworthy political and societal disturbance. However, the healthcare improvements have been powerful assets in mitigating the impacts.

1.2 Definition Of Pandemics

Pandemic is derived from two Greek words: *pan* meaning "all" and *demos* meaning "the individuals" [2]. A pandemic is a wide-reaching spread of a novel disease. A pandemic happens when a novel infection shows up and spreads the world over and the vast majority does not have the obligatory immunity. Infections that have instigated past pandemics normally originated from animals. Studies suggest that pandemic strains experience vital genomic transmutations called *antigenic shift*. For WHO to articulate a level-VI pandemic strain alert, there must be a sustained spread in minimum two regions simultaneously.

Pandemics have been extensively utilized to define ailments that are novel or are at least related to the novel alternatives of existing viruses. Yet, its relative idea expresses that seven *cholera pandemics* occurred during the past 200 years, presumably, all triggered by alternates of same virus. Pandemics are *non-infectious* or *non-transmissible* such as *obesity risk behavior* or *infectious* and *transmissible* like *SARS* and *COVID-19*.

1.3 History Of Pandemics

A portion of the history's most destructive pandemics beginning from *Antonine Plague* to the *COVID-19* are as per the following [3, 4].

1.3.1 Prehistoric Epidemic

a. Circa: 3000 BC
   Around 5000 years back, an epidemic cleared out an ancient town in China. Skeletons of infants and middle-aged individuals were discovered inside the houses. All age groups were affected and the preserved archaeological location is now called *"Hamin Mangha."*
b. Plague of Athens: 430 BC
   After the war amid Athens and Sparta, an epidemic emaciated Athens for almost 5 years. The symptoms of the epidemic were debatable, yet some suggested it like a typhoid fever.
c. Antonine Plague: 165–180 AD
   It was believed that the epidemic was brought after the war between Roman and Parthia – the Roman Empire to the end of Roman Peace Period. More than 5 million people lost their lives.
d. Plague of Cyprian: 250–271 AD
   It was described as the signal for the end of the world. It was projected to have killed 5000 people per day in Rome. Mass burial sites were found by the archaeologists in 2014 in Rome.
e. Plague of Justinian: 541–542 AD
   This plague occurred in the Byzantine empire and ravaged it completely and was one of the prominent reasons for its downfall. The plague is known after emperor Justinian who constructed a great cathedral *"Hagiya Sophia"* in Constantinople, modern day Istanbul, which was a museum till date but now is being converted into a mosque.
f. The Black Death: 1346–1353
   It traveled from Asia to Europe causing havoc all over. Some suggested that it had wiped out half of Europe population. Mass graves were

recovered later. It changed Europe history, and studies have suggested that survivors had better access to bread and meat. Deaths caused scarcity of labors which forced industries to undergo technological innovation.

g. American Plagues: 16th Century

It was brought to America by European explorers. 90% of indigenous population of America was wiped out due to it, which helped in some way for the unopposed exploration.

h. Great Plague of London: 1665–1666

Fleas from plague infected rodents were the cause of its transmission. About 100,000 people, which included 15% of the population of London, had died.

i. Russian Plague: 1770–1772

It was one of the deadliest plagues in the history of Russia. It wreaked havoc, otherwise the greatest empire, and ravaged as many as 100,000 people were killed.

j. Flu Pandemic: 1889–1890

Transport travel in the industrial age made its spread easy. Though it started in St. Petersburg, it spread across the whole Europe, despite no air travel.

1.3.2 Modern Epidemics

a. American Polio Epidemic: 1916

It started in New York City affecting mostly children and, at times, leaving survivors with permanent disabilities. It occurred sporadically in USA until the development of *Salk* vaccine in 1954. In India, vaccination against Polio started in 1978. In 2014, WHO declared India a polio-free country.

b. Spanish Flu: 1918–1920

It is one of the deadliest pandemics in the modern history. It spread during the World War I. Like the symptoms of modern-day coronavirus, it spread widely especially among soldiers due to their poor and unhygienic living conditions [5].

c. Asian flu: 1957–1958

It was influenza like and started in China. The disease claimed more than 1,000,000 lives all over the world.

d. AIDS 1981 – Present Day

It most likely started in West Africa in the 1920s. The virus made its way around the world and has claimed more than 35,000,000 lives since

it was identified. As of now, the disease has no known cure, but with regular treatment, people experience normal life.
  e. H1N1 Swine Flu 2009–2010
    Swine flu pandemic was caused by a new strain of H1N1 which originated in Mexico. It mostly affected younger age group; the virus infected as many as 1.4 billion people [6].
  f. EBOLA 2014–2016
    It ravaged West Africa between 2014 and 2016 and the first case was reported in Guinea. There is no cure for EBOLA till date [7].
  g. Zika Virus: 2015
    It spread through mosquitoes of *Aedes* genus, but in humans, it can also be sexually transmitted. It is mostly harmful for infants and it flourished in warm humid climate, making South America and Central America the prime areas for it.

1.4 Attributes Of A Pandemic

Despite the fact that the expression "pandemic" has not been characterized by many clinical writings, there are some key characteristics of a pandemic, which help us to comprehend it better.
  a. Wide Geographic Extension
    Pandemic typically alludes to sicknesses that stretch out over huge geographic zones, for instance, The Black Death, cholera, flu, AIDS, etc. In a categorical inspection of the chronicles of pandemic flu, pandemics were sorted as trans-territorial and global [8]. There were 178 nations affected by the H1N1 in 2009.
  b. Disease Movement
    In extension to geographic augmentation, most usages of the expression pandemic suggest surprising sickness development or through transmission trackable here and there. Instances of illness development incorporate man-to-man spread of infections through respiratory infections like flu and SARS, or by vectors, for instance, dengue.
  c. Novelty
    Pandemics have been exploited usually to depict illnesses that are novel, or if nothing else related with new alterations of prevailing viruses, for instance, antigenic mutations occurring in *influenza*, the commencement of HIV/AIDS in the mid-1980s, and documented pestilences of pandemics, for example, plague. "There have been 7 cholera pandemics

during the previous 200 years, apparently completely brought about by variations of a similar living being" [9].

d. High Mortality Rates and Explosiveness

Pandemics are portrayed by high mortality ratios and by hazardous widespread, e.g., Ebola. On the contrary, if the transmission is not explosive in nature, irrespective of its spread across the board, it may not be termed pandemic, e.g., West Nile infection. Low paces of indicative ailment are rarely delegated pandemics, in any event, even when they spread generally.

e. Minimal Population Immunity

Even though pandemics frequently have been portrayed in partly resistant populaces, obviously in restricting microbial contamination and transmission, populace immunity can be a ground-breaking tool against pandemic power. Pandemics are characterized by populace immunity; henceforth, it is simple for a huge piece of populace to be contaminated.

f. Infectious and Contagious

Pandemic is usually not employed to portray non-infectious maladies, for instance, obesity, or hazardous customs, like smoking cigarettes, that are topographically extensive and might be expanding to a worldwide scenario and, however, are not contagious. Pandemics are contagious and this transmission can be direct (man-to-man) or indirect (man-vector-individual). It is like the SARS was communicated from man-to-man, while H7N9 was spread by contacting poultry. The significance to human wellbeing of this avian flu strain can be accessed by its capability to convert into a form equipped for persistent man-to-man transmission.

1.5 Origin Of The Coronavirus Or Covid-19

In late December, news broke of a mysterious fever in China and connecting regions like Taiwan and Hong Kong. The primary known human contaminations in China were accounted for in December. Human to human transmission was affirmed by WHO on 20 January 2020. Many of these cases were connected to Hunan sea food market, Wuhan, China, which sold wild animals. China said that the *vet* market was only the spreading site not the episode site. There were a few affirmed hypotheses that the hints of infection were found in the sewage waters of urban communities of Italy in mid-December, and some detailed of its beginning in the USA. In any case, till date, the wellspring of the source or *patient zero* could not be established; some alleged that a 55-year-old shrimp dealer in Wuhan was the first ever case. However, it ended up being a mere speculation with no official affirmation. By late December, the transmission got human to human in China and was expanding at a progressive pace and was affirmed by Chinese CDC as another coronavirus having pneumonia-like manifestations to some degree like SARS from obscure causes. On 30 January 2020, WHO proclaimed the Corona infection as the general wellbeing crisis of global worry; at this point, the outbreak spread by 100 to multiple times.

Since late January, countries from Europe and the USA started reporting the infection which has now spread across all over the world. As of August 2020, more than 18 million cases have been reported, resulting in more than 700,000 deaths.

In January 2020, WHO recommended the name *2019 nCOV* or state NOVEL CORONAVIRUS, an acute respiratory illness. Finally, in February, the official names COVID-19 and SARS-CoV-2 were issued. This infection is accepted to have a *zoonotic* origin. It hereditarily bunches with the genus Betacoronavirus, 95% indistinguishable from bat coronavirus samples at the genome level, so is accepted to have been moved from bats to people via an intermediate host have which is again alleged as Pangolin.

1.5.1 Pathophysiology

COVID-19 influences, for the most part, the upper respiratory tracts (sinuses, nose, and throat) and the lower respiratory tracts (windpipe and lungs). Lungs are the most affected as the infection gets to the host cells by means of the ACE2 (angiotensin changing over catalyst) which is generally plentiful in lungs; the infection utilizes surface called *glycoprotein*, a spike, to associate with ACE2 and enter the host cell. As the alveolar ailment advances, respiratory failures happen, which prompts casualty. It can likewise cause respiratory failures by attacking the central nervous system (CNS). It additionally influences the gastrointestinal organs as ACE2 is richly present in the cells of gastric and duodenal cells of the *small intestine*. It can likewise harm the cardiovascular framework as ACE2 are profoundly engaged with the heart work, vein dysfunctions, and clot formation; complications with kidneys are some other affected dimensions.

1.5.2 Signs, Symptoms, and Transmission

COVID-19 is an entirely novel corona virus; so its subtleties are still under investigation. In any case, the accessible examinations show that it spreads between individuals through close contacts by coughing, sneezing, and talking or standing closely. The small droplets fall on a superficial level, and individuals contacting the surface and afterward contacting their faces make them contaminated. Transmission can likewise happen through *aerosols*, little droplets that remain suspended in the air for noticeably long occasions. Episodes have been accounted for in crowded and inadequately ventilated insides like workplaces and cafés and night clubs. The infection can be distinguished for up to four hours on copper, up to one day on cardboard, and up to three days on plastic and stainless steel. Sputum and saliva likewise also convey infections. Fever is the most widely recognized side effect; different indications incorporate cough, loss of appetite, fatigue, muscle and joint pain, cold, vomiting, diarrhea, shortness of breath, which are the most critical symptoms. The moment an individual is infected and the time the individual builds up a symptom is called an *incubation* period, which varies from 5–14 days. Numerous cases do not show indications and are called *asymptomatic carriers* who communicate contamination, without demonstrating any side effects, and a large portion of the enlisted cases are asymptomatic. Also, there are *presymptomatic carriers* who do not feel sick but will eventually display symptoms. Finally, there are *symptomatic carriers* who show all the indications.

1.5.3 Diagnosis

WHO has published several testing protocols for COVID-19. The most reliable is RT-PCR (real-time reverse transcription polymerase chain reaction). The test is done on respiratory samples acquired by a nasopharyngeal swab; results from this test are available from few hours to two days. *Antibody test*, otherwise called the serological tests, is done from blood samples which can recognize active infections and whether the individual has been contaminated before; yet, its exactness is still faulty around 60%–70%. Also, *chest CT scan* is tested in India which is useful in diagnosing the individuals with the doubt of contaminations. TruNat machines are likewise utilized in India, a diagnostic machine for tuberculosis, for COVID-19; they help in testing the drug resistant TB.

1.5.4 Prevention

No vaccine for COVID-19 is yet accessible nor are there any recommendations for any drug for avoidance of COVID-19. With no prophylactic measures, the only way could be by postponing the pandemic pinnacle, known as the "Flattening of Curve" which could be accomplished by slowing the contamination and taking into consideration better health facilities. Preventive measures suggested by WHO and CDC are as per the following:

- Wearing face cover openly is mandatory, as keeping up physical separation is outlandish out in the open spots and there is an expanded danger of transmission, face veils limit the volume and travel distance of the inhaled and exhaled droplets.
- Keeping up cleanliness after a cough and sneeze ought to be encouraged, and washing hands with cleanser and water normally, particularly before contacting the nose, face, eyes, or eating food, ought to be promoted.
- CDC has additionally suggested alcohol-based hand sanitizers with at least 60% alcohol to be used. Sodium hypochlorite and bleach mixture is utilized in plenitude for sanitization.
- Social distancing ought to likewise be urged to decrease contacts with infected individuals or enormous gatherings where transmission could be effectively conceivable by shutting down or closing educational institutions, gyms, malls, cinema halls, etc., and canceling large public gatherings like marriages and other functions.
- Work from home is additionally being supported; online meetings and classes are likewise every now and again conducted.

1.5.5 Management

The individuals who have been determined to have COVID-19 are encouraged to isolate themselves and look for clinical assistance. Infected patients are being managed with symptom treatments like fluid therapy, oxygen support, and supporting other affected vital organs. Hygiene, healthy lifestyle, and diet have been prescribed to improve immunity. Since the severity fluctuates with various strains, some may get few or no symptoms. Mellow cases recoup inside up to 14 days while those with severe or critical cases may take over three weeks. Risk in kids is lower when contrasted with adults, grown-ups over 50 years old, and with comorbidity circumstances like hypertension, diabetes, and heart have graver dangers. Also, smoking and air contamination and unhygienic chores add to an expanded hazard. A few investigations have proposed that men have a death rate of 2.8% and women have a death rate of 1.7%; however, as we state, all the exploration and studies are in progress.

1.6 Types Of Covid-19

There are six different types of COVID-19 infections suggested by the recent studies. Sampling different kinds of symptoms could help determine the actual likelihood of COVID-19 turning dangerous and patient undergoing hospitalization.

a. Cluster I

It is the mildest form of infection due to the trouble in the upper respiratory tract brought on by an increased viral load. People suffering from this infection experience symptoms such as cold, sore throat, blocked chest pain, muscle pain, loss of smell, and headache, but fever was absent in this phase.

b. Cluster II

It is a little more cumbersome than the Cluster I. Patients belonging to this category have symptoms of a mild flu-like infection persisting fever and a loss of appetite.

c. Cluster III

Patients suffering from this experience impact on their digestion and gastrointestinal functioning. Cough was not a prominent symptom in this cluster. Nausea, loss of appetite, vomiting, and diarrhea were commonly observed; headaches and chest pain were also observed.

d. Cluster IV

The symptoms observed in this cluster are related to energy loss and exhaustion brought on by immunity slow down. Considered as a severe cluster, patients in this category have symptoms like fatigue, headache, loss of smell and taste, sore throat, fever, and chest pain.

e. Cluster V

This is severe Level II, with confusion, more severe than Level I; the type of symptoms in this cluster impacted nervous system functioning and is the start of the lasting impact that COVID could possibly have on the brain. Headache, loss of smell, appetite, cough, fever, confusion, sore throat, fatigue, and muscle pain are some of the commonly observed symptoms.

f. Cluster VI

It is severe Level III, with abdominal and respiratory distress. This is the most alarming and severe kind of symptoms that are seen in people in the first weeks. Other than experiencing the common symptoms like confusion, sore throat, shortness of breath, muscle, and abdominal pain, people in this cluster are more likely to undergo hospitalization and require ventilation and oxygen support. Identifying these clusters helps in the determination of severity of the infection and later in the treatment.

So, as we discuss about treatment, symptomatic treatments from COVID-19 are being given till date due to the lack of a vaccine. Talking about vaccine, more than 150 countries are engaged in the race for the COVID-19 vaccine to save and bring normalcy in the world.

1.7 Vaccine

It is a biotechnology proposed to provide gained immunity against the infection. For COVID-19, as of August 2020, 231 vaccine candidates are in development, albeit none has finished the clinical preliminaries to demonstrate its safety and viability. In August, 24 vaccine candidates were declared for undertaking clinical preliminaries with six beginning Phase III. In February 2020, WHO was not expecting a vaccine against SARS-CoV-2 to become available under a year and a half. Past ventures to create vaccines for viruses in the family *Coronaviridae* like SARS and MERS have just been tested on non-human animal models. But no vaccine or cure has been demonstrated to be benign and effective in humans. Since the detection of COVID-19 in December 2019, its genetic sequence was printed on 11 January 2020 prompting an imperative retort to prepare for the outbreak and fasten the production of a preventive drug. As of today, America's *Moderna*, Oxford's *Astrozenca* and *Sinovac* from China are ahead in the race against time for COVID-19.

Many a times, vaccine trials fail; in the case of COVID-19 specifically, a vaccine efficacy of 70% may be enough to stop the pandemic, but a 60% efficacy may lead to the continuation of the outbreak.

1.8 Pandemic Impacts

Infectious outbreaks can cross borders and threaten economic and regional stability. Some of the after-effects are as follows [10].

a. Health Effects

Pandemics infect millions, causing widespread illness and deaths. During the SARS outbreak in China in 2003, more than 8000 people were infected and 700 died. Influenza pandemic has a high transmission and fertility rate with 250,000–500,000 in 2004. In Mexico, May 2009 H1N1 virus capable of man-to-man contiguity had witnessed around 2000 deaths in 180 countries. In USA from 2009 to 2010, H1N1 virus infected 43–89 million cases with 8000–19,000 deaths approximately. *Spanish Flu* in 1919–1920 caused 20–40 million deaths, while *Asian Flu* in 1957–1958 caused 2 million deaths. Other major threats in recent times were *Dengue* and *Ebola*. In 2015, *Ebola* outbreak was witnessed in West Africa with around 12,000 deaths. Similarly, in 2015–2016, *Dengue* epidemics were felt in Latin America witnessing 1.5 million cases. And very recently, in 2020, it is the COVID-19 that is wreaking havoc with more than 20 million cases and nearly 10 lakh deaths all over the world.

Though many people have recovered, losing one life for any family is horrendous. COVID-19 infection does not give lasting immunity against the disease after recovery. So, chances of getting infected are very high. Recent studies have revealed that even after recovery, people experience other health issues like blood clotting, kidney damage, cardiovascular problems, lungs damage, and many more. Many mental health problems have also been experienced.

b. Economic Impacts

Pandemics create lasting effect not only on health but also on livelihood and can create a crippling effect and instability in the society. The cost of dealing with a pandemic could be very high. It is like the Ebola outbreak that crippled the West African economies. The Global Health Risk Framework for the Future (GHRF) estimated that, every year, the infectious disease outbreaks cost 60 billion USD as direct costs. The cost of hospitals, staff, and medication soar up during a pandemic crisis like today in the COVID-19 world. There are urgent needs for ventilators, medication, safety equipment like PPE kits, hand gloves, face masks, sanitizers, upgradation in the hospitals, and arrangement for beds. As India has the world's largest COVID-19 center in Delhi, it does put a strain on the government expenditure and indirectly hampers the developments. The indirect effects of pandemics include decline in the GDP. For example, SARS outbreak resulted in the decrease of 1% of GDP of China. The introduction of COVID-19 containment measures across the world made GDP in the G-20 nations fall by 3.4% in the first quarter of 2020, which is the largest contraction since 1998. GDP fell by only 1.5% during the first quarter in 2009 at the height of financial crisis. [11]. Among the G-20 economics, those that introduced stringent lockdown measures at the earliest saw the largest contractions in the GDP in the first quarter of 2020 like China (−9.8%), USA (−1.3%), UK (−20%), and Japan (−0.6%). Some other after-effects due to the lockdown and restricted travel bans were as follows:

   i. Sharp rise in unemployment due to closing of schools, colleges, hotels, factories, and many more sectors like these.

   ii. Stress on the supply chains due to closing of the factories and business, which gave rise to unemployment and less liquidity with the people, so less purchasing which lead to a decrease in government income.

   iii. Tremendous collapse of tourism and hospitality industry.

iv. Pecuniary turmoil related with the present pandemics has widespread and severe implications upon budgetary business sectors, including stocks bonds and commodities like crude oil and gold. The UNDP expects a US $220 billion drop in revenue in developing countries and its impact to last for many upcoming months.

v. When emergencies strike, ladies are diligently hit by the monetary effects as women hold more unreliable positions in the informal and service sector with lesser social protection. This leaves them less ready to ingest the economic setbacks.

vi. And last but not least, the aviation industry has been hit the hardest due to travel restrictions.[12].

So, we can assert that pandemics have not only immediate but also long-term effects on the economy.

c. Social Impacts

Pandemics have had a far-reaching consequence beyond the spread of the disease itself, including political, cultural, and social implications as we talk of the present pandemic COVID-19 and effects of quarantining [13], which are as follows:

i. Political Impacts

Many governments were criticized for its handling of the outbreak, like the CPC of China; US president Donald Trump was criticized for his misleading information and for downplaying the pandemic's significance. It has worsened the conflict dynamics across the world and has emphasized the potential erosion of political and economic sovereignty, somehow paving the way for an entirely new geo-political scenario.

ii. Social Implications

It has affected the educational system to a great extent, leading to closing of schools and colleges [14]. It has affected the religious gatherings and worships due to closing of mosques, churches, and temples. But mental health was worst affected and there were psychosocial issues around the world. There is the potential spike in suicides exacerbated by social isolation, unemployment, and financial factors. Financial crisis has led to increased aggression at home and domestic violence and intimate partner violence. Older persons living in nursing homes are vulnerable to infection and those living alone have been hit very hard as they face barriers in obtaining food, medication, essential supplies, etc. The impacts of the pandemic have affected the life around the world in every aspect possible.

d. Security Impacts

A security peril by pandemic is unquestionably a reiterating occurrence. Worldwide security is compromised from pandemics, undoubtedly. Pandemics are not, at this point, restricted to only the domain of general prosperity and clinical prescription, but to a social subject, a progressive issue, and a global security issue as well. Bioterrorism including weaponries and bioterrorist assaults are frequently originated from the normally occurring and mutating contagious disease upsurges, as the conduct of security has changed remarkably, particularly over the recent decades, veering off a long way from the more customary security. Governments need to zero in additionally on the military availability to defy the antagonistic impact of plagues and pandemics capability [15–18].

1.9 Conclusion

A pandemic incorporates a wide geographic augmentation illness development, seriousness, insignificant populace immunity, contagiousness, and infectiousness. Each pandemic is unique, which makes anticipating its repercussions more of an educated guess than science. There are very few models that contrast with the most pessimistic scenario assessments of something like COVID-19; the closest correlation in current times occurred over a century back – the *Spanish Flu* which attacked the globe from 1918–1919.

The pandemic-related calamity has caused colossal negative effects on the wellbeing, economies, and even national security. The contrary effects are intense; the pandemic has tainted a great many individuals, causing far-reaching ailments and a large number of causalities. It represents a genuine danger not exclusively to the populace, however, economies also. Financial misfortune can bring about unsteadiness of the economy through immediate and roundabout expenses. The social effects are significantly more genuine because of the movement bans, education system, market closure, business limitations, etc.

A productive compelling and co-operative reaction can diminish the mortality, monetary, and social effects. It will be a basic task of the administrations to manage a pandemic outbreak in the present and future.

References

[1] MacKellar, L. (2007). Pandemic influenza: a review. Population and Development Review, 33(3), 429-451.

[2] MacKellarSource:, L. (2007). Pandenic Influenza: A Review. Vol. 33, 24.

[3] Potter, C. W. (2001). A history of influenza. *Journal of applied microbiology, 91*(4), 572-579.

[4] LePan, N. (2020). Visualizing the history of pandemics. *Visualizing the History of Pandemics.*

[5] Lin, L., McCloud, R. F., Bigman, C. A., & Viswanath, K. (2016). Tuning in and catching on? Examining the relationship between pandemic communication and awareness and knowledge of MERS in the USA.

[6] Rewar, S., Mirdha, D., & Rewar, P. (2015). Treatment and Prevention of Pandemic H1N1 Influenza. Annals of Global Health, 81(5), 645-653.

[7] Gostin, L. O., Tomori, O., Wibulpolprasert, S., Jha, A. K., Frenk, J., Moon, S., . . . Dzau, V. J. (2016). Toward a Common Secure Future: Four Global Commissions in the Wake of Ebola. PLoS Med, 13(5), e1002042

[8] Rewar, S., Mirdha, D., & Rewar, P. (2015). Treatment and Prevention of Pandemic H1N1 Influenza. Annals of Global Health, 81(5), 645-653.

[9] Honigsbaum, M. (2009). Historical keyword Pandemic. The Lancet, 373.

[10] Morens, D. M., Folkers, G. K., & Fauci, A. S. (2009). What is a pandemic? J Infect Dis, 200(7), 1018-1021. doi: 10.1086/644537

[11] Creighton, C. (1965). *A history of epidemics in Britain* (Vol. 1).

[12] Hays, J. N. (2005). *Epidemics and pandemics: their impacts on human history.*

[13] Chung, L. H. (2015). Impact of pandemic control over airport economics: Reconciling public health with airport business through a streamlined approach in pandemic control. Journal of Air Transport Management, 44, 42-53.

[14] Chen, W.-C., Huang, A. S., Chuang, J.-H., Chiu, C.-C., & Kuo, H.-S. (2011). Social and economic impact of school closure resulting from pandemic influenza A/H1N1. Journal of Infection, 62(3), 200-203.

[15] Davies, S. E. (2013a). NATIONAL SECURITY AND PANDEMICS. UN Chronicle, 50(2), 20-24.

[16] Davies, S. E. (2013b). National Security and Pandemics. UN Chronicle, 50, 20-24.

[17] Rosenberg, C. E. (1992). *Explaining epidemics.* Cambridge University Press.

[18] Bourdelais, P. (2006). *Epidemics laid low: a history of what happened in rich countries.* JHU Press.

2

Tracing the Origins of COVID-19

Vidushee Nautiyal[1] **and Rakhi Pandey**[2]

[1]Department of Education, Uttaranchal College of Education, Dehradun,
Uttarakhand 248001, India
[2]Department of Humanities and Social Sciences, Jaypee University,
Anoopshahr, Uttar Pradesh 203390, India
Corresponding Author: Vidushee Nautiyal, vidusheenautiyal11@gmail.com.

Abstract

Human race has defeated and survived a number of pandemics and epidemics
for thousands of decades. Battle for survival has been there in the genes of the
human race; no matter what, they surpassed the danger. But standing on the
verge of destruction has made the humans helpless to defend them currently.
In the series of such pandemics, coronavirus disease has once again gained
prominence amongst virologists and the medical practitioners across the
globe. Scientists and doctors from all over the world are trying to unravel the
mystery of coronavirus; yet, it is obscure. The world is still burying corpses
due to this artificial virus emergence. To summarize knowledge about the
ongoing virus and situation, we have undertaken an analysis of the publicly
accessible literature and data. In this literature review, we tried to explore the
history of pathogenic viruses and the origin of COVID-19.

2.1 Introduction

According to the Oxford dictionary, a virus is an infectious agent that
typically consists of a molecule of nucleic acid in a protein shell, is far too
modest to get through to a light microscope, and can only multiply with in a

living host cell. Viruses are named on the grounds of their genetic structure to enable the production of diagnostic tests, vaccines, and drugs as the largest and the best scientific community does. On 31 December 2019, the World Health Organization (WHO) Office, China was notified that clusters of pneumonia of an unknown cause were being recorded in the city of Wuhan, one of the densely populated cities, within the Hubei Province of China. China authorities identified this as a kind of coronavirus (COVID-19) that was previously unknown. Since the initial outbreak, COVID-19 number of suspected cases has been increasing rapidly across China and worldwide. WHO has declared the disease as a pandemic and Public Health Emergency of International Concern (PHEIC). Originated from animal-to-human transmission in the wholesale seafood and live animal meat market of Wuhan, has not only gobbled the people of China but also through person-to-person transmission has swallowed millions of people worldwide. The study shows that the virus is common in certain species of animals, where the bats show the closest genetic similarities while pangolins can also be the major cause of the origin. On the other hand, there is an accusation that the virus was prepared in the laboratory as a bioweapon by China to strike down the world economy and create its own power.

As of 29 August 2020, the number of COVID-19 cases confirmed is 24,587,513, with 833,556 deaths according to WHO. Among all the countries, 5,811,519 cases were confirmed from the United States of America, and there were 179,716 deaths cumulative total. In India alone, the number of confirmed cases are 3,463,972 and 62,550 deaths have been enlisted with WHO. Table 2.1 lists the total number of COVID-19 positive cases across different regions of the world.

Flipping back to the pages of zoonotic virus (a virus that is regularly transmitted from an animal to a human), we can find the traces of the birth of Severe Acute Respiratory Syndrome Coronavirus 2 (SARS-CoV-2) from the preceding viruses (SARS) and Middle Eastern respiratory syndrome virus (MERS). Coronaviruses were first detected during the 1930s when chronic respiratory disease of trained chickens triggered an inevitable bronchitis

Table 2.1   COVID-19 report by WHO.

Country	Total Cases	Status
America	13,005,995	Confirmed
Europe	4,180,463	Confirmed
South-East Asia	3,987,855	Confirmed
Eastern Mediterranean	1,891,458	Confirmed
Africa	1,038,418	Confirmed
Western Pacific	482,583	Confirmed

Table 2.2 Classification of corona virus.

Alpha coronavirus	Human coronavirus (HCoV)-229E and HCOV-NL63
Beta coronavirus	HCOV-0C43, Severe Acute Respiratory Syndrome Coronavirus, and Middle Eastern Respiratory Syndrome Virus
Gamma coronavirus	Virus of whales and birds
Delta coronavirus	Virus of pigs and birds

infection (IBV) [1]. During the 1940s, two additional creature coronaviruses, mouse hepatitis infection (MHV) and the Contagious Gastroenteritis Infection (TGEV), came into existence [2]. But, at that time, no one realized that these three viruses were interrelated. In the 1960s, human coronavirus was discovered [3, 4]; since then, a lot of theories and discoveries came into existence and got demolished, and even the meaning of coronavirus which was earlier in 1960s got changed to what we call coronavirus today. Essentially, coronaviruses are the group of similar RNA viruses that has root diseases in mammals and birds. This virus has a positive single-stranded RNA which belongs to the kingdom: Orthornavirae; class: Pisoniviricetes; family: Coronaviridae. The Coronaviridae subfamily is divided into four major subdivisions [5]: the Alpha coronavirus, the Beta coronavirus, the Gama coronavirus, and the Delta coronavirus as shown in Table 2.2. The Alpha and the Beta coronaviruses infect mammals and the Gamma and Delta coronaviruses chiefly infect birds [6] as the Alpha coronavirus includes the Human Coronavirus (HCoV)-229E and HCoV-NL63, the Beta coronavirus has HCoV-0C43, Severe Acute Respiratory Syndrome Human Coronavirus (SARS-HCoV), HCoV-HKU1 and the MERS-CoV. The Gamma coronavirus has the viruses of whales and birds and the Delta coronavirus has viruses isolated from pigs and birds [7]. COVID-19 or SARS-CoV-2 belongs to the same gene family of the initial pathogenic viruses, SARS-CoV and MERS-CoV.

The paper organization is as follows. Section 2.2 introduces the history of viruses and Section 2.3 elucidates the genetic sequence of SARS-CoV-2. Section 2.4 discusses the transmission of the virus. Section 2.5 depicts how the virus has been diagnosed and Section 2.6 provides conclusive remarks to summarize the paper.

2.2 History of the Virus

It is common to identify the latest disease with some other outbreaks in the recent history with the fresh cases of the current coronavirus disease, COVID-19, rising day after day. For example, the influenza of 1918 that affected almost one-third of the world's population before it fizzled away. Certain alarming viruses which emerged out of the box were: SARS, H1N1 Influenza in 2009, and Ebola. The effect of each disease is determined primarily on factors – when did it originate, where did it originate, how is it transmitted, contagiousness and lethalness of the virus, how hygienic people are, what precautions they are taking, and how soon a vaccine or cure is available. Both SARS-CoV-2 secluded from humans to date are closely genetically similar to the corona virus secluded from the poor population, especially the bats of the subdivision Rhinolophus. SARS-CoV, which triggered the SARS pandemic in 2003, is also closely similar to the corona virus isolated from birds. Bats of Rhinolophus genus are spread across Africa, Asia, the Middle East, and Europe. SARS-COV-2 is not present in farmed or domestic animals as being genetically linked to other identified coronavirus. Analysis of the sequences of the virus genomes also shows that SARS-COV-2 is very well-suited to receptors of human cells, thereby allowing it to enter human cells and easily contaminate people. It is not only the death rate or the number of people who are being affected that depicts how deadly the disease is, but it is even the effect on the economy, the day-to-day life of people, and the whole ecological system.

2.2.1 Influenza

Affecting almost one-third of the world's entire population in 1918, Spanish flu epidemic became the deadliest flu of the season. With a whole lot of different lifestyles, living in a crowded area, struggling in the battlefields, having less knowledge of vaccines and medicines, and less advancement of laboratories and science, people faced a lot of problems to cure the disease. The only option they had was to lie down to dead beds. It is possible to classify complete data architecture in four classes, Tier from I to IV, each one having some advantages and disadvantages associated with power usage and accessibility. The availability and reliability concerns yield to redundant $N+1$, $N+2$, or 2N data center designs and this has a significant impact on power usage.

With the major symptoms of fever, nausea, aches, and diarrhea, it was first reported in March 1918 affecting 500 million people and more than 50 million deaths, 675,00 in the USA itself, and the mortality rate was about 2% nationwide. It was transmitted by breathing droplets that influenced the community of people between the ages of 20 and 40 years. There was no medication available for the outbreak; there were no antibiotics or antivirals, and not even the vaccines. The pandemic ended in the summer of 1919 largely because of debts and higher levels of immunity.

2.2.2 Seasonal Flu

Every year, the virus hits but no two seasons are identical. Since strains evolve each year, predicting what will strike can be difficult. Unlike COVID-19, we have safe vaccinations and antiviral drugs that can aid in the prevention and reduction of flu severity. The major problem is that the human body is somewhat used to the previous flu; on the contrary, COVID-19 is new for the human body to adjust with.

With the main symptoms of fever, cough, sore throat, and fatigue, up to 5,000,000 of which were serious were 9% of the population or around 1 billion infections. The death rate ranged from 291,000 to 646,000 wearing about 0.1%. It transmitted through its respiratory droplets where this was being targeted by older adults and people with compromised immune systems. Numerous antiviral medicines such as Tamiflu, Relenza, Rapivab, and Xofluza have been used to minimize the length and severity of the flu. Many vaccine options were available which provided immunity against multiple influenza strains.

2.2.3 2002–2004: Severe Acute Respiratory Syndrome

In 2002, the world witnessed the deadliest virus of the 21st century, reported as SARS. Primarily, it came into recognition in the second last month of the year 2002. People in Guangdong province of China complained about a disease similar to an influenza, starting with a migraine and myalgia and proceeding to fever, and later elevated to severe atypical pneumonia, respiratory failure, and then death [8]. The first origin of the SARS-CoV was suspected in the exotic animals in the market place in Guangdong and later got transmitted to humans during both 2002 and 2003 also extending to the winters of 2004. The Himalayan palm civets (*Paguma larvata*) and the raccoon dogs (*Nyctereutes procyonoides*) fostered SARS-CoV-like diseases. The Himalayan Palm civet is of particular concern; as the virus may be removed from most markets, SARS-CoV can survive for weeks in palm civets. In addition, intermittent infections noticed in 2003–2004 were related to restaurants where they cooked and consumed palm civet meat [8]. It was proved that bats turned out to be the prime source of the virus SARS-CoV. The species of bats, especially the horseshoe bats (subdivision – Rhinolophus), was considered the reason for its diversity. The bat being the host of the virus in the absence of unconcealed disease lasted in the genetic diverseness of the virus.

It was first identified in Guangdong province of China in November 2002, with the main symptoms of fever, respiratory symptoms, cough, and malaise.

There were 774 deaths in 29 countries with global cases of 8098 with 15% mortality rate. It transmitted droplets and infected surfaces across the respiratory tract and consumed people older than 60. There was no therapy or cure, but certain people were operating on antiviral drugs and steroids. By the time the pandemic ended in July 2003, a vaccine was available.

2.2.4 2009 (H1N1) Flu Pandemic

The (H1N1) Flu pandemic resulted as one of the most suffered pandemics for nine months beginning from January 2009 and lasted till August 2010. The researchers believed that the virus resided several months before the outbreak at Mount Sinai School of Medicine. The virus was present in latent pigs, as a desperate increase in agriculture observation in order to avoid potential upsurge [9]. Originally identified as an "outbreak," the famous infection was primarily reported in the states of Mexico, proving its activation several months before its actual announcement. Like other influenza strains, the H1N1/09 pandemic virus did not affect adults over 60 years of age disproportionately, which was the rare feature of the H1N1 infection.

Having symptoms of fever, chills, cough, and body aches, it was first identified in Mexico in January 2009 and about 24% of the global population that was targeted in the United States in April 2009, with more than 284,000 deaths and a death rate of 0.02%. Children have the highest risk of being affected, and 47% of children aged between 5 and 19 years experienced the symptoms compared to 11% of people aged 65 and over. Antivirals such as Oseltamivir and Zanamivir were available to help many people recover without any problem; the hunt for H1N1 vaccine began in April 2009 and ended in December 2009. The pandemic finally ended in August 2010.

2.2.5 Middle East Respiratory Syndrome Coronavirus (MERS-COV) – 2012

For the first time, the disease was identified in Saudi Arabia in the month of April 2012; over 2400 cases of coronavirus respiratory syndrome (MERS-CoV) in the Middle East have been reported in 27 countries. Confirmed cases which have been identified in Europe had direct or indirect contact with the Middle East. The spring of the virus remained uncertain, but the patterns of transmission and the virological studies pointed to Dromedary camels of the Middle East as a pool from which the humans were irregularly infected by zoonotic transmission [10]. The virus was first reported in Saudi Arabia in 2012 with the main symptoms of fever, cough, shortness of breath, pneumonia, and gastrointestinal symptoms. It was a non-human to human transmission that transmitted the main reserve oil host of MERS via the dromedary camel, and this is the source of infection in humans. Thirty-five percent of total infected patients died as a result of the virus – it was mainly males over 60 years of age with underlying medical conditions such as diabetes, hypertension, and renal failure. There is still no vaccine or advanced treatment available for MERS today.

2.2.6 2014–2016 Ebola

The WHO has declared Ebola a global emergency. Due to its rapidly increasing and devastating effect in the Democratic Republic of Congo. As per Dr. Shanthi Kappagoda, who is an infectious disease specialist at Stanford Health Care, an Ebola outbreak generally occurs when someone comes in an exposure with an infected animal, like a primate or a bat, specifically by slaughtering, cooking, or eating it. It spreads through human to human transmission when the infected person's body fluid such as sweat, urine, semen, breast milk, feces, and blood comes in contact with another.

December 2013 with main symptoms of fever aches and pain exhaustion as well as diarrhea and vomiting and the first outbreak occurred in March 2014. There were 28,652 cases in 10 countries with 11,325 deaths and a death rate of about 50%. It transmitted by body fluids such as sweat, blood and parts, near contact and became most infectious toward disease finishing. There was no such medication in 20% of all cases in infants; only vaccines were available, but supportive care was given, including IV fluids only oral rehydration. The epidemic was brought to an end in March 2016.

2.3 Genetic Sequence of SARS-CoV-2

The corona virus genome, ranging in length from 26 to 32 kg base, contains a number of open reading frames (ORFs). There have been claims that the SARS-COV-2 genomes have 14 ORFs encoding the 27 proteins [10, 11]. The exterior spikes of glycoprotein play a very important role in joining the host cell with the recipient and are vital for deciding host tropism and communication ability, mediating receptor binding and membrane bonding [12]. The virus has a diameter ranging from 60 to 140 mm, has protein spikes involved, and the overall structure has genetic material that looks identical to other Coronaviridae family viruses [13]. On comparison of SARS-CoV-2 and SARS-CoV, researchers discovered that they were quite similar, but the proteins present had some notable difference. On the other hand, when

SARS-CoV-2 was juxtaposed with MERS-CoV, they found that it was less related. SARS-CoV-2 and SARS-CoV have a similar genome pattern which is linked to the genes of bats (Rhinolophus). Overall, at genome stage, shares of SARS-CoV-2 have 87.99% pattern identity with the bat-SL-COVZC45 and 87.23% pattern identity with the bat-SL-COVZXC2, which is slightly genetically identical to SARS-CoV which is almost 79% and MERS-CoV which is almost 50% [14].

This study clearly shows that the bats have the most diverse corona viruses found in the bodies and happen to be the host of other types of coronavirus, like the MERS-CoV and the SARS-CoV [15]. Bats, being the host of the virus, transmitted the virus to humans through the Himalayan Palm Civet in SARS-CoV and through the Dromedary Camels in MERS-CoV, but, till date, the intermediate host has not been found for SARS-COV-2. There are various reasons that prove bats to be the original host of the virus but not as the intermediate host [16]:

a. In the wet seafood livestock wholesale market of Wuhan, a lot of non-aquatic animals along with the mammals were present for purchase, but no trace of bats was found.
b. The close relative bats to SARS-CoV-2, SL-CoVZC45, and the bat SL-CoVZXC21 have long branches, suggesting that they are not directly related to SARS-CoV-2.
c. In other coronaviruses, such as SARS-CoV and MERS-CoV, there has been an intermediate host for transmitting the virus from bats to humans.

Research by scientists is somewhat directed to the pangolins' coronavirus genomes having 85.5%–92.4% to be the in-between host responsible for the SARS-CoV-2 [16]. With all these evidences, it is clear that SARS-CoV-2 has a natural animal origin and the accusation which was viral in the social media through the news agencies against China holds no importance and evidence to prove it. It is neither a manipulated or constructed virus nor any bioweapon created by China to strike down the world economy [17].

2.4 Transmission and Diagnosis

In the last quarter of the year, December 2019, with the first few medical cases filed on 31 December, reporting the problem of pneumonia, from the wet livestock selling meat market of Wuhan, China were discovered, making Wuhan as the birthplace for SARS-CoV-2. Unable to resolve the problem, Wuhan health commission immediately reported to the WHO about the severe attack of the virus in the public. Many of the SARS-CoV-2 initial patients had direct contact with Wuhan's livestock market; they were the stall owners, the market workers, or the regular market visitors. Environment samples collected from the market in December 2019 tested positive for COVID-19, indicating that the markets were the source of the outbreak or played a role in the initial magnification of the outbreak [18]. The cases which occurred in late December are clearly directed toward Wuhan to be the epicenter of the virus, but the cases which were determined on 1 December 2019 had no concern with the livestock sea-food retailing of Wuhan. Genetically, this virus was launched into the market from the other one but a secret location, where it spread speedily, which means all human to human transmissions must have occurred previously [18, 19]. Possibilities suggest that the unrecognized infection would have occurred in the mid-November 2019 which was around 14–15 days prior, when the latest infectants of the virus were discovered. As of 25 August 2020, 23,951,003 are live cases and 820,007 are death toll across 213 countries and territories. Due to this rapid elevation of positive reports across the globe, the health organization, on 11 March 2020, declared the coronavirus as the pandemic and warned the public about the worst which was yet to come. As for the confirmed cases of coronavirus, the USA, Brazil, India, Russia, and Peru are the five countries most affected. The animal–human transmission and the human–human transmission is said to occur among close contacts because the respiratory droplets are formed when an infected person coughs or sneezes. SARS-CoV lives up to 96 hours on surfaces. The other coronavirus is up to nine days [20, 21]. Hubei Province in central China was shut down on 23 January with all movements blocked in and out of the Province. Traveling through China was discouraged and the number of available scheduled flights and train journeys dramatically decreased to only 10% of previous operations. Commercial and social events were marginal, and restaurants and other entertainment spots along with schools and most shops were also shut down.

The seventh edition of the pneumonia diagnosis and treatment program for the novel coronavirus infection, as reported by the National Health Commission of the People's Republic of China, classified doubtful cases as persons with fever or respiratory problems. Confirmed cases of COVID-19 were identified as a positive result of high-performance sequencing or reverse transcription polymerase chain reaction (RT-PCR) testing of respiratory specimens including nasal and pharyngeal swab specimens, bronchoalveolar lavage fluid, sputum or bronchial aspiration, or as a positive result of anti-SARS-CoV-2 IgM/IgG or anti-SARS-CoV-2 IgG antibody for times higher recovery [22]. COVID-19 is being diagnosed through clinical characteristics, chest imaging, epidemiological history, and through laboratory detection.

2.5 Conclusion

The emerging pandemic, COVID-19, is obviously the world health problem. What one knows about the pathogen is the way in which it affects the cells and then causes various diseases, and the clinical features of diseases have been making rapid progress. Because of the fast transmission, countries around the world have decided that attention should be given to disease observation systems and country preparedness and reaction should be increased. COVID-19 plundered 60 countries at once which shared the possibility of health and humane disaster, which affected three out of four children due to inadequate hand washing service at their classrooms. The worldwide school functioning system has narrowed the consumption of water (drinking and non-drinking). The various health reports proved that approximately 700 million children suffered deterioration in their general hygiene when in school. UNICEF and WHO are drawing the attention of the government to execute a proper sanitization procedure for a hygienic environment. This will not only help in controlling the spread and protecting the children from infection but also be beneficial for other forthcoming infections. Hence, it would provide them with a carefree hygienic learning.

2.6 Acknowledgment

The authors are thankful to anonymous reviewers and editors for their suggestions.

References

[1] T. Estola, "Coronaviruses, a new group of animal RNA viruses", Avian diseases, pp. 330–336, 1970.

[2] K. McIntosh, "Coronaviruses: a comparative review", In Current Topics in Microbiology and Immunology/Ergebnisse der Mikrobiologie und Immunitätsforschung, Springer, Berlin, Heidelberg, pp. 85–129, 1974.

[3] J. S. Kahn and K. McIntosh, "History and recent advances in coronavirus discovery", The Pediatric infectious disease journal, Vol. 24, No. 11, pp. S223-S227, 2005.

[4] E. Mahase, Covid-19: First coronavirus was described in The BMJ in 1965, 2020.

[5] F. Li, "Structure, function, and evolution of coronavirus spike proteins", Annual review of virology, Vol. 3, pp. 237–261, 2016.

[6] Q. Tang, Y. Song, M. Shi, Y. Cheng, W. Zhang and X. Q. Xia, "Inferring the hosts of coronavirus using dual statistical models based on nucleotide composition", Scientific reports, 5, 17155, 2015.

[7] C. J. Burrell, C. R. Howard and F. A. Murphy, "Fenner and White's Medical Virology", Academic Press, 2016.

[8] W. Li, S. K. Wong, F. Li, J. H. Kuhn, I. C. Huang, H. Choe and M. Farzan, "Animal origins of the severe acute respiratory syndrome coronavirus: insight from ACE2-S-protein interactions", Journal of virology, Vol. 80, No. 9, pp. 4211–4219, 2006.

[9] G. J. Smith, D. Vijaykrishna, J. Bahl, S. J. Lycett, M. Worobey, O. G. Pybus, ... & J. M. Peiris, "Origins and evolutionary genomics of the 2009 swine-origin H1N1 influenza A epidemic", Nature, Vol. 459, No. 7250, 1122–1125, 2009.

[10] Z. Song, Y. Xu, L. Bao, et al., From SARS to MERS, thrusting coronaviruses into the spotlight, Viruses 11, http://doi.org/10.3390/v11010059.

[11] A. Wu, Y. Peng, B. Huang, X. Ding, X. Wang, P. Niu, ... & J. Sheng, "Genome composition and divergence of the novel coronavirus (2019-nCoV) originating in China", Cell host & microbe, 2020.

[12] Z. Zhu, Z. Zhang, W. Chen, Z. Cai, X. Ge, H. Zhu, ... & Y. Peng, "Predicting the receptor-binding domain usage of the coronavirus based on kmer frequency on spike protein", Infection, Genetics and Evolution, 61, 183, 2018.

[13] B. Udugama, P. Kadhiresan, H. N.Kozlowski, A. Malekjahani, M. Osborne, V. Y. Li, ... and W. C. Chan, "Diagnosing COVID-19: the disease and tools for detection", ACS nano, Vol. 14, No. 4, pp. 3822–3835, 2020.

[14] R. Lu, X. Zhao, J. Li, P. Niu, B. Yang, H. Wu, ... Y. Bi, "Genomic characterisation and epidemiology of 2019 novel coronavirus: implications for virus origins and receptor binding", The Lancet, 395(10224), pp. 565–574, 2020.

[15] J. Cui, F. Li, and Z. L. Shi, "Origin and evolution of pathogenic coronaviruses", Nature Reviews Microbiology, vol. 17, no. 3, pp. 181–192, 2019.

[16] H. Harapan, N. Itoh, A. Yufika, W. Winardi, S. Keam, H. Te, ... & M. Mudatsir, "Coronavirus disease 2019 (COVID-19): A literature review", Journal of Infection and Public Health, 2020.

[17] T. T. Y. Lam, N. Jia, Y. W. Zhang, M. H. H. Shum, M. H. H., J. F. Jiang, H. C. Zhu, ... and W. J. Li, "Identifying SARS-CoV-2-related coronaviruses in Malayan pangolins", Nature, 1–4, 2020.

[18] Origin of SARS-Cov-2, World Health Organization, 2020.

[19] W. B. Yu, G. D. Tang, L. Zhang and R. T. Corlett, "Decoding the evolution and transmissions of the novel pneumonia coronavirus (SARS-CoV-2/ HCoV-19) using whole genomic data", Zoological Research, pp. 41, no. 3, pp. 247, 2020.

[20] A. Kramer, I. Schwebke and G. Kampf, "How long do nosocomial pathogens persist on inanimate surfaces? A systematic review", BMC infectious diseases, Vol. 6, No. 1, pp. 130, 2006.

[21] G. Kampf, D. Todt, S. Pfaender and E. Steinmann, "Persistence of coronaviruses on inanimate surfaces and their inactivation with biocidal agents", Journal of Hospital Infection, Vol. 104, No. 3, pp. 246–251, 2020.

[22] National Health Commission of the People's Republic of China. New coronavirus pneumonia prevention and control program (trial version 7th ed).http://www.nhc.gov.cn/yzygj/ s7653p/202003/46c9294a7dfe4cef80dc7f5912eb1989.shtmlAccessed 4 Apr 2020

3

AI for COVID-19: The Journey So Far

Abhinav Sharma[1], Arpit Jain[1], and Mangey Ram[2]

[1]Department of Electrical and Electronics Engineering, University of
Petroleum and Energy Studies, Dehradun, Uttarakhand, 248007, India
[2]Department of Mathematics, Graphic Era University, Dehradun,
Uttarakhand 248002, India
Corresponding Author: Arpit Jain, arpit.eic@gmail.com.

Abstract

COVID-19 is an infectious virus caused by acute respiratory syndrome
SARS-CoV-2. On 11 March, the World Health Organization (WHO)
declared COVID-19 as pandemic with around 118,000 cases across 100
countries. Being a highly infectious disease, it has raised significant
challenges to global health systems. With the sustained risk of global
spread, the health care industry is looking forward to new technologies
to fight against this disease. Artificial intelligence (AI), Internet of Things
(IoT), blockchain technology, robotics, etc., are some of the smart and
innovative technologies which have revolutionized health care industry in
recent past. In this chapter, we will focus on the role of AI technology which
includes machine learning (ML), deep learning (DL), and computer vision
(CV) techniques in monitoring and controlling the spread of COVID-19.
ML technique is widely being explored for predicting the spread of this
virus and also provides useful information to control its spread. CV
techniques are useful to diagnose the infected patients through computed
tomography (CT) and magnetic resonance imaging (MRI). Drug industry
is also exploring AI for rapid design and development of COVID-19
vaccine. AI enabled robots are used in hospitals to decrease the

workload of doctors and medical staff. In near future, this technology will emerge as a useful tool to fight against other disruptions affecting the life of human being.

3.1 Introduction

COVID-19 has hit globally at a colossal scale. With worldwide reported cases of 28.329 million and being declared as pandemic by World Health Organization (WHO), it has led to severe impact on humanity. Being a highly contagious disease, it has put up global health services to its severe challenges. This is the high time for research community around the world to gather up and provide their contribution in fighting this crisis. The health care professionals are battling at the frontend by contributing relentlessly, with the biological and health care researchers being deeply involved in the vaccine/treatment development. This pandemic has united the entire mankind for developing innovative solutions to fight the pandemic. There are numerous ways in which modern technologies can assist the mankind for fighting against the pandemic; some of them are: population screening, contact tracing, robots assisting health care professionals, minimizing human contact with autonomous systems, speeding up the drug discovery, efficient diagnostics, and chatbots for various sectors, to name a few. Various countries are fighting to minimize the losses due to the outbreak; however, a common trait is enforcing lockdown, which has become the main defense mechanism for India as well. At the time of writing, India had more than 0.9 million [1] active cases, amidst a population of 1.3 billion. This has presented an interesting case of epidemic fight to the world when the anticipated value of R0 for COVID-19 cases lies between 2 and 3 [2]. However, a major sacrifice made to flatten the curve is the Indian economy, which has suffered due to continuous lockdown from 23 March 2020. India reported its first case of COVID-19 on 30 January 2020 with infection origin from China [3]. Tracking the first 50 cases of COVID-19 in India gives us a big picture of state-wise spread and the steps endured by the government to handle the epidemic and their efficiency [4]. On 30 January, India reported its first three COVID-19 cases in the state of Kerala. This was followed by Delhi, Telangana, and Tamil Nadu. The situation became more critical when the state of Rajasthan reported 17 cases (16 Italian tourists along with their 1 Indian driver). On 5 March, the national capital region of India (NCR) reported cases in Ghaziabad and Gurgaon. Kerala reported the youngest patient on 9 March, a 3-year-old girl who had a travel history to Italy. On 10 March, Maharashtra's Pune reported two cases, the first two cases for the state. The first case of third-level transmission was discovered in Meerut, a person who came in contact with six patients in Agra, who contracted the infection from a patient in Delhi who had travel history to Italy. As of 12 September, the number of cases has increased to 4.5 million in India. This paper presents the journey of artificial intelligence (AI) in predicting and controlling the spread of COVID-19 across the world. In March 2020, smart hospitals were built in Wuhan, China for COVID-19 patients who use AI and Internet of Things (IoT). Hospitals use smart IoT devices for contactless health monitoring of patients and robots for cleaning and sanitizing different areas of hospitals.

The manuscript is outlined in the following manner. Section 3.2 presents the brief overview of AI techniques. Section 3.3 presents the potential contribution of AI against COVID-19. Section 3.4 provides the conclusive remarks to summarize the paper.

3.2 Artificial Intelligence

The word AI means automation of intelligent behavior and was first coined in the year 1956 at the Dartmouth conference, USA. AI is the subset of computer science and is defined as the science and engineering of making intelligent machines, especially intelligent computer programs. Alan Turing [5] and John McCarthy are the two scientists who will be remembered for the discovery of AI. Since its inception, the technology has experienced exponential growth in different sectors of our day-to-day life. With the availability of higher end embedded system and 70 years of research in AI programming, our dreams are becoming a reality where humans can think of doing countless feats that were never possible before. IoT is another booming technology which basically collects, sends, and acts on data that it collects from the environment. AI and IoT, part of industrial revolution 4.0, have jointly revolutionized health, agriculture, and automation industry.

AI is an interdisciplinary field and has diverse subfields, and each field in itself is an innovative area of research. Machine learning (ML), deep learning (DL), expert system (ES), natural language processing (NLP), artificial neural network (ANN), soft computing, knowledge representation and planning, game playing, fuzzy logic (FL), and computer vision (CV) are the subareas of AI as shown in Figure 3.1. ML gives the computer the ability to learn without being explicitly programmed. In ML, machines learn, adapt, and improve with a set of data without human assistance. ML algorithms are classified into three categories, i.e., supervised, unsupervised, and reinforcement learning algorithms. Supervised learning algorithms work on labeled dataset, unsupervised learning algorithms work with

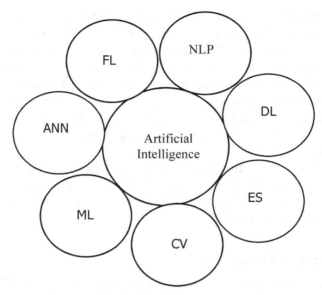

Figure 3.1   Subareas of artificial intelligence.

unlabeled dataset, while reinforcement learning algorithms learn from the environment through reward and punishment. DL is a subset of ML and explores hierarchical structure of ANN to learn and generalize in a non-linear fashion. ML and DL algorithms are widely explored by e-commerce and entertainment sites such as Amazon, Flipkart, and Netflix for identifying the user taste. Health care industries are exploring the potential of ML and DL algorithms for drug discovery and predict its success rate based on human biological factors. CV techniques, another vital subarea of AI, basically train the computer to interpret and understand the visual world. It uses DL models to identify and classify objects in digital image dataset. Medical fraternity is using CV techniques for disease identification and classification.

ES is a computer program that exhibits intelligent behavior and solves real-world problems through reasoning from knowledge-based system. The system uses the database of expert knowledge and gives suggestions or helps in making decision in diverse areas of science and engineering such as medical diagnosis, fault identification in vehicles, process control system, etc. FL also exhibits the decision-making capability of humans, which involves all intermediate possibilities to deal with uncertain problems in biomedical engineering. NLP is one of the important fields of AI that enables computers to understand and generate human speech. NLP uses ML algorithms to extract meaning from human language. Virtual assistant applications, Google duplex, Google translator, and robots use NLP to understand syntax and semantics of unstructured language. NLP-based intelligent chatbots are used by a number of multi-national companies to provide customer services to end users. Health care industry is using AI powered NLP to assist doctors in dictating observations and automate filling of health record. All these AI techniques are playing a significant role in developing and upgrading the medical system on a global scale.

3.3 Potential Contribution of AI Against Covid-19

AI is an emerging technology in the area of health care industry and has actively contributed to the battle against COVID-19. BlueDot [6] and MetaBiota [7], two AI-based companies, predicted the outbreak of this infectious disease before its declaration as a pandemic by WHO. AI-based ML and DL techniques coupled with radiological imaging provided accurate diagnosis of this disease and eliminates the problem of limited number of specialized medical staff and reverse transcription polymerase chain reaction (RT-PCR) test kits in remote villages. AI and computational drug designing are enlightening new paths for repurposing the existing drug as well as discovery of new drug for combating COVID-19. ML- and DL-based prediction techniques are widely explored for predicting the survival and mortality rate of COVID-19 positive patients. Once the human is diagnosed with COVID-19, then the next important action is to prevent its spread. Many countries are utilizing AI-based contact tracing applications to break the virus transmission chain. New generation of AI-based macro- to micro-scale robots have revolutionized the medical industry and had played an important role by assisting medical fraternity in this pandemic situation. Researchers have explored ML model to estimate the transmission route of Middle East Respiratory Syndrome (MERS) coronavirus. Many leading private and public organizations are struggling to build communication with their customers due to enforced lockdown; therefore, giant companies are deploying AI-based chatbots to provide validated information about their products and spread awareness about the ongoing pandemic. The only measure to prevent the further spread of this disease is social distancing. AI-based CV techniques find useful applications to monitor crowded areas and reduce the probability of contagion.

3.3.1 Diagnosis of Disease

Intelligent medical imaging has played a vital role in the diagnosis of human diseases such as breast cancer detection, pneumonia detection, lungs and brain disease classification, to name a few. In the starting phase of COVID-19 pandemic, medical fraternity in China diagnosed the virus using computed tomography (CT) and chest X-ray images due to the availability of limited test kits. AI-based ML and DL predictive models provided good assistance to radiologist for accurate, cheaper, and faster diagnosis and prognosis of disease. In [8], the author proposed COVIDX-Net DL network which comprises seven convolutional neural networks (CNNs) for the detection of COVID-19 using X-ray images. In [9], the author explored COVID-Net, a deep CNN for the diagnosis of disease using 13,975 X-ray image datasets, and simulation results show that the proposed model presents accuracy, sensitivity, and positive predictive value (PPV) of 93.3%, 91.0%, and 98.9%. In [10], the author utilized data available on GitHub, Kaggle, and Open-i to diagnose COVID-19 from X-ray images using ResNet50 CNN model with support vector machine (SVM) algorithm. The simulation results show that the proposed classification model outperforms other models with an accuracy of 95.38%. In [11], the author proposed DarkNet CNN architecture with 17 convolutional layers having different filtering schemes on each layer. The proposed architecture accurately provides binary classification (COVID vs. No-Findings) with an accuracy of 98.08% and multi-class classification (COVID vs. No-Findings vs. Pneumonia) with an accuracy of 87.02%. The study also concluded that the proposed architecture can also be used for diagnosing other chest diseases like tuberculosis and pneumonia.

In [12], the author proposed 24 layers nCOVnet CNN model for diagnosing the disease in less than 5 seconds with an accuracy of 97.62%. In [13], the author proposed robust sustainable Bayes-SqueezeNet CNN classification model that hybridizes 15 layers SqueezeNet architecture with Bayesian optimization algorithm. The proposed model identifies normal, pneumonia and COVID cases from X-ray image dataset with an accuracy of 98.3%. In [14], the author proposed DeepCOVIDExplainer for detecting COVID-19 symptoms from chest X-ray images. The author explored VGG (Visual Geometry Group), ResNet (Residual Networks), and DenseNet CNN (Densely connected Convolutional Network) architecture and the simulation results show that DenseNet-161 outperforms other models with precision, recall, and F1 score of 95.2%, 94.5%, and 94.8% for balanced dataset and 89.3%, 87.4%, and 88.3% for unbalanced dataset. In [15], the author explored random forest (RF) ML model to classify COVID patients from chest CT image dataset comprising 1658 images of positive cases and 1027 negative cases. The proposed model presents accurate results with sensitivity, specificity, and accuracy of 90.7%, 83.3%, and 87.9%. In [16], the author presents COVID-MobileXpert which is a deep neural network based mobile application that can detect COVID-19 from noisy chest X-ray image.

Once the patient is diagnosed with COVID-19, then, based on the level of infection, medical resources can be provided to the patients. It has been observed that infants and elderly people who have weak immunity level are most affected from this disease. In [17], the author explored ML algorithm to predict the disease severity level from database of blood samples obtained from Wuhan, China so as to provide on-time medical assistance to infected people and reduce the mortality rate. In [18], the author utilizes decision trees (DT), RF, and SVM ML predictive models to predict patients at high risk and may go on to develop acute respiratory distress syndrome (ARDS). The proposed model was trained with dataset obtained from two hospitals in China which presents an accuracy of 80% to predict severe cases.

3.3.2 Discovery of Drug and Vaccine

AI technology is potentially contributing in the discovery of new drugs at a much faster rate as well as in the clinical trials during development of the vaccine. Baidu researchers developed AI-based predictive algorithm LinearFold that can predict SARS-CoV-2 RNA (ribonucleic acid) structure [19]. Google's DeepMind predicted the protein structure of the virus through DL-based AlpaFold system. Developing the 3D model of protein structure is helpful in understanding how the virus functions and finally in the development of its vaccine [20]. In [21], the author explored DL model to detect old drug with anticoronavirus activities by learning the model with two datasets of 3C-like protease constraint and other data-holding records of infected SARS-CoV, SARS-CoV-2, influenza, and human immunodeficiency virus (HIV)). The study suggested that eight drugs, bedaquiline, brequinar, celecoxib, clofazimine, conivaptan, gemcitabine, tolcapone, and vismodegib, are effective against the feline infectious peritonitis (FIP) corona virus. In [22], the author utilized DL model namely Molecule Transformer-Drug Target Interaction (MT-DTI) to detect commercially available drug that can prevent from viral proteins of SARS-CoV-2. The simulation results showed that a drug antiretroviral, namely atazanavir, which is widely used to prevent HIV, is the best chemical compound that can fight against COVID-19. Benevolent AI, Innoplexus, Deargen, Gero, Cyclica, Healx, and VantAI are some of the pharmaceutical companies that are repurposing the existing drug to fight against COVID-19, while Insilico Medicine, Exscientia, Iktos, and SRI international are some other pharmaceutical companies that are actively involved in discovering new drug for COVID-19 [23]. Although AI is providing fast and cost-effective methodology for drug development, clinical trials are required for final validation.

3.3.3 Prediction of Mortality and Survival Rate

Prediction of the spread of virus and providing the guidelines or prevention measures is another AI application in COVID-19. Kaggle and GitHub are the two websites where the real-time data of COVID-19 segregated based on confirmed cases, active cases, cured cases, and deaths of each country and state are available. This dataset can be used for predicting the active cases across different regions of the world so that appropriate amount of health infrastructure can be made available to these places. In [24], the author predicted the transmission dynamics of COVID-19 across different regions of China using auto-encoder and clustering algorithm. A dataset between 11 January 2020 and 27 February 2020 was collected from WHO and the total number of confirmed cases and transmission dynamics were identified with good accuracy.

With limited number of medical resources such as ventilators, it is important to identify mild patients with potential malignant progression. In [25], the author explored logistic regression and DL model for accurately predicting patients with malignant progression using clinical data and chest CT images. In [26], the author explored logistic regression model to identify clinical characteristics of COVID-19 patients and build risk score for predicting intensive care unit (ICU) admission and mortality of patients. The performance of the proposed model was evaluated using area under the curve (AUC) and the results show that AUC of 0.74 and 0.83 was obtained for predicting ICU admission and mortality rate. In [27], the author utilized three clinical parameters, lactic dehydrogenase (LDH), lymphocyte, and high-sensitivity C-reactive protein (hsCRP) to predict the survival rate of critical patients with more than 90% accuracy using XGBoost algorithm. In [28], the author explored XGBoost ML model that can predict biomarkers of disease mortality 10 days before with an accuracy of more than 90%. A dataset of 485 patients which includes pregnant women and patients less than 18 years old were collected between 10 January 2020 and 18 February 2020 from Wuhan, China. The study showed that patients having high level of LDH require immediate medical treatment to reduce the mortality rate.

3.3.4 Contact Tracing

COVID-19 is one of the most infectious viruses experienced on earth as compared to other viruses. WHO reported that the virus primarily spread from person to person; therefore, contact tracing is the main solution to stop further spread of virus. Contact tracing means to detect the person who recently came in contact of COVID-19 patient. Manual contact tracing through interviews conducted by medical authorities does not always provide right information because, many times, COVID-19 patients come in contact with unknown person in supermarket, airports, etc. Therefore, it became a challenging task for government authorities to do the manual contact tracing.

In this respect, researchers have come up with digital contact tracing applications in smartphones that can identify the COVID-19 patients through Bluetooth, global positioning system (GPS), and network-based API [29]. These applications automate the contact tracing process and are built using centralized, decentralized, and hybrid architecture. These virtual applications also collect the human data and analyze it through ML and DL model to detect whether the person is infected with virus due to their recent contact with COVID-19 patients [30]. The success of digital technologies in contact tracing will be guaranteed through the voluntary adoption of these applications in smart devices.

3.3.5 Robotics and Health Care

In the last few years, robots have played a significant role in health care industry [31]. Robots are filling the health care gaps and are providing innovative solutions to medical fraternity. They are assisting doctors in surgical procedures and are also reducing the risk of medical staff by examining the patient with contagious disease [32]. AI-assisted intelligent humanoid robots find applications such as telemedicine, handling contaminated waste, health monitoring of patients, and decontamination. China has explored many of these applications in hospitals to reduce the spread of COVID-19. In Italy, Tommy named robots have been used for measuring blood pressure, oxygen saturation, and temperature of the COVID-19 patients [33]. COVID-19 not only spreads from person to person but also through contaminated surface; therefore, robots also find applications in disinfecting and sterilizing the public places.

In the current pandemic situation, many countries have faced a challenge of lack of qualified doctors and nurses to take test samples and process them. Robots have been widely adopted for sample collection, transfer, and testing, thereby reducing the risk of infection and workload of medical staff [34]. Robots are also used for food and medicine delivery in hospitals as well as entertaining patients in hospitals and quarantine centers so as to maintain patient's mental health. Automated drones can find application to monitor the movement of people in a denser city under lockdown and also find applications in sanitizing the affected areas [35].

3.3.6 COVID-19 Chatbots

Chatbots are AI-based virtual assistant application that can built interaction between human and computer in natural language. In the current pandemic situation, chatbots are widely explored in health care to spread awareness and handle queries of millions of people across the world [36]. WHO and Centre for Disease Control and Prevention (CDC) have already incorporated chatbots in their websites to provide instant information about COVID-19. WHO launched chatbot on Facebook Messenger to educate people about this disease in English, Spanish, French, and Arabic languages. The Government of India has also launched chatbot to handle queries related to COVID-19. Apart from health care industry, these chatbots are also adopted by police control rooms and other public sectors directly and indirectly involved in this pandemic. Chatbots provide a wide range of information such as symptoms of disease, transmission route of infection, preventive measures, health and travel advisories, and official government helpline numbers for further assistance.

In [37], the author designed a chatbot utilizing GPT-2 language model for answering queries related to COVID-19. The quality of generated responses were improved using Term Frequency-Inverse Document Frequency (TF-IDF), Bidirectional Encoder Representations from Transformers (BERT), Bidirectional Encoder Representations from Transformers for Biomedical Text Mining (BioBERT), and Universal Sentence Encoder (USE) to filter and retain relevant sentences in the responses. The simulation results show that BERT and BioBERT outperform TF-IDF and USE in filtering of sentences. In [38], the author introduced COVINFO reporter chatbot developed on a commercially available chatbot platform Quirobot to provide crucial pandemic information to the general public. The chatbot was designed, implemented, and evaluated between March and April 2020.

In [39], the author developed an AI-based conversational bot "Aapka Chikitsak" on the Google Cloud Platform (GCP) to bridge the gap between patients and clinical staff in the current pandemic situation. This virtual assistant chatbot provides telehealth services like preventive measures, counseling sessions, health care information, and symptoms of COVID-19 with an intuitive interface. India is a country with diverse religion, and people speak different languages. Health care chatbots that have been developed in the past do not provide multi-lingual virtual assistant support to people living in rural areas. This chatbot provide multi-lingual health support (Hindi and English) to diverse population of India. Symptoma is a symptom-to-disease virtual health assistant chatbot which can differentiate more than 20,000 diseases with an accuracy of more than 90%. In [40], the author evaluated the performance of Symptoma with a set of diverse clinical cases and results show that the chatbot can accurately distinguish COVID-19 cases with an accuracy of 96.32%.

3.3.7 Prevent Further Spread of Disease

WHO reports and national and local public health authorities have clearly specified that till the development of COVID-19 vaccine, the only measure that can prevent the further spread of this disease is to use surgical masks and follow the social distancing guidelines. AI-based CV techniques that combines image processing with ML and DL techniques find useful applications to recognize people who are not wearing a surgical mask. China has used infrared cameras in public places such as airports and railway stations to scan people with high temperature. Thereafter, the identified infected people are quarantined. Some countries have used AI-based face recognition applications that recognize those individuals who are not following the self-quarantine rules. ML and DL models can be used to predict the number of cases in a particular area and can help the government agencies to maintain a specific number of health care professionals and resources in that area. Therefore, AI provides a preventive measure to fight and prevent the further spread of this disease.

3.4 Conclusion

AI is a growing technology and has shown tremendous potential to fight against COVID-19 and other similar pandemics. ML and DL algorithms are assisting health care professionals in diagnosis and prognosis of this disease. Government and other health authorities are also using these predictive models for contact tracing, screening, and predicting the spread of this disease. AI-based robots are used in hospitals for monitoring the health of infected people and sanitizing the infected areas. Unmanned aerial vehicle (UAV) with thermal imaging cameras are used to scan the public places in order to ensure that social distancing is properly maintained. Microsoft Bing's COVID-19 Tracker, UpCode, and NextStrain are some of the dashboards that have been created to visualize the actual and expected infected cases. Drug companies are exploring AI techniques to develop drug and vaccine for this contagious disease. Although AI is playing an important role, still it is at the preliminary stage and will take time to completely overcome this pandemic situation.

References

[1] https://www.worldometers.info/coronavirus/country/india/

[2] M. D'Arienzo and A. Coniglio, "Assessment of the SARS-CoV-2 basic reproduction number,R0, based on theearly phase of COVID-19 outbreak in Italy," Biosafety and Health, pp. 1–3, 2020.

[3] A. Tomar and N. Gupta, "Prediction for the spread of COVID-19 in India and effectiveness of preventive measures," Science of The Total Environment, vol. 728, 2020.

[4] M. Rawat, "Coronavirus in India: Tracking country's first 50 COVID-19 cases; what numbers tell," 12 March 2020. [Online]. Available: https://www.indiatoday.in/india/story/coronavirus-in-india-tracking-country-s-first-50-covid-19-cases-what-numbers-tell-1654468-2020-03-12. [Accessed 1 June 2020].

[5] TURING, I. B. A. (1950). Computing machinery and intelligence-AM Turing. Mind, 59(236), 433.

[6] Niiler, E. An AI Epidemiologist Sent the First Warnings of the Wuhan Virus. Available online: https: //www.wired.com/story/ai-epidemiologist-wuhan-public-health-warnings/ (accessed on 29 January 2020).

[7] Heilweil, R. How AI Is Battling the Coronavirus Outbreak. Available online: https://www.vox.com/recode/2020/1/28/21110902/artificial-intelligence-ai-coronavirus-wuhan (accessed on 29 January 2020).

[8] E.E.D. Hemdan, M.A. Shouman, M.E. Karar, COVIDX-Net: A Framework of Deep Learning Classifiers to Diagnose COVID-19 in X-Ray Images, 2020 arXiv preprint arXiv:2003.11055.

[9] Wang, L. and Wong, A. (2020). COVID-Net: A Tailored Deep Convolutional Neural Network Design for Detection of COVID-19 Cases from Chest Radiography Images. arXiv, 22 March(https://arxiv.org/abs/2003.09871).

[10] Sethy, P. K., & Behera, S. K. (2020). Detection of coronavirus disease (covid-19) based on deep features. Preprints, 2020030300, 2020.

[11] Ozturk, T., Talo, M., Yildirim, E. A., Baloglu, U. B., Yildirim, O., & Acharya, U. R. (2020). Automated detection of COVID-19 cases using deep neural networks with X-ray images. Computers in Biology and Medicine, 103792.

[12] Panwar, H., Gupta, P. K., Siddiqui, M. K., Morales-Menendez, R., & Singh, V. (2020). Application of Deep Learning for Fast Detection of COVID-19 in X-Rays using nCOVnet. Chaos, Solitons & Fractals, 109944.

[13] Ucar, F., & Korkmaz, D. (2020). COVIDiagnosis-Net: Deep Bayes-SqueezeNet based Diagnostic of the Coronavirus Disease 2019 (COVID-19) from X-Ray Images. Medical Hypotheses, 109761.

[14] Karim, M. R., Döhmen, T., Rebholz-Schuhmann, D., Decker, S., Cochez, M., & Beyan, O. DeepCOVIDExplainer: Explainable COVID-19 Diagnosis Based on Chest X-ray Images.

[15] Shi, F., Xia, L., Shan, F., Wu, D., Wei, Y., Yuan, H., ... & Shen, D. (2020). Large-scale screening of covid-19 from community acquired pneumonia using infection size-aware classification. arXiv preprint arXiv:2003.09860.

[16] Li, X., Li, C., & Zhu, D. (2020). COVID-MobileXpert: On-Device COVID-19 Screening using Snapshots of Chest X-Ray. arXiv preprint arXiv:2004.03042.

[17] Yan, L., Zhang, H. T., Goncalves, J., Xiao, Y., Wang, M., Guo, Y., ... & Huang, X. (2020). A machine learning-based model for survival prediction in patients with severe COVID-19 infection. MedRxiv.

[18] Jiang, X., Coffee, M., Bari, A., Wang, J., Jiang, X., Huang, J., ... & Wu, Z. (2020). Towards an artificial intelligence framework for data-driven prediction of coronavirus clinical severity. CMC: Computers, Materials & Continua, 63, 537–51.

[19] Huang, L., Zhang, H., Deng, D., Zhao, K., Liu, K., Hendrix, D. A., & Mathews, D. H. (2019). LinearFold: linear-time approximate RNA folding by 5'-to-3'dynamic programming and beam search. Bioinformatics, 35(14), i295-i304.

[20] https://syncedreview.com/2020/03/05/google-deepmind-releases-structure-predictions-for-coronavirus-linked-proteins/.

[21] Ke, Y. Y., Peng, T. T., Yeh, T. K., Huang, W. Z., Chang, S. E., Wu, S. H., ... & Lin, W. H. (2020). Artificial intelligence approach fighting COVID-19 with repurposing drugs. Biomedical Journal.

[22] Beck, B. R., Shin, B., Choi, Y., Park, S., & Kang, K. (2020). Predicting commercially available antiviral drugs that may act on the novel coronavirus (SARS-CoV-2) through a drug-target interaction deep learning model. Computational and structural biotechnology journal.

[23] Kaushik, A. C., & Raj, U. (2020). AI-driven drug discovery: A boon against COVID- 19?. AI Open, 1, 1–4.

[24] Hu, Z., Ge, Q., Jin, L., & Xiong, M. (2020). Artificial intelligence forecasting of covid-19 in china. arXiv preprint arXiv:2002.07112.

[25] Bai, X., Fang, C., Zhou, Y., Bai, S., Liu, Z., Xia, L., ... & Xie, X. (2020). Predicting COVID-19 malignant progression with AI techniques.

[26] Zhao, Z., Chen, A., Hou, W., Graham, J. M., Li, H., Richman, P. S., ... & Duong, T. Q. (2020). Prediction model and risk scores of ICU admission and mortality in COVID-19. PloS one, 15(7), e0236618.

[27] Yan, L., Zhang, H. T., Xiao, Y., Wang, M., Sun, C., Liang, J., ... & Tang, X. (2020). Prediction of survival for severe Covid-19 patients with three clinical features: development of a machine learning-based prognostic model with clinical data in Wuhan. medRxiv.

[28] Yan, L., Zhang, H. T., Goncalves, J., Xiao, Y., Wang, M., Guo, Y., ... & Huang, X. (2020). An interpretable mortality prediction model for COVID-19 patients. Nature Machine Intelligence, 1–6.

[29] Ahmed, N., Michelin, R. A., Xue, W., Ruj, S., Malaney, R., Kanhere, S. S., ... & Jha, S. K. (2020). A survey of covid-19 contact tracing apps. IEEE Access, 8, 134577-134601.

[30] Lalmuanawma, S., Hussain, J., & Chhakchhuak, L. (2020). Applications of machine learning and artificial intelligence for Covid-19 (SARS-CoV-2) pandemic: A review. Chaos, Solitons & Fractals, 110059.

[31] Alotaibi, M., & Yamin, M. (2019, March). Role of Robots in Healthcare Management. In 2019 6th International Conference on Computing for Sustainable Global Development (INDIACom) (pp. 1311–1314). IEEE.

[32] B. S. Shivaji, J. A. Mushtaqahmed, and G. U. Suresh, "MEDICAL ROBOTICS AND AUTOMATION," vol. 2, no. 4, p. 9, 2013.

[33] Romero, M. E. (2020, April 08). Tommy the robot nurse helps Italian doctors care for COVID-19 patients. https://www.pri.org/stories/2020-04-08/tommy-robot-nurse-helps-italian-doctors-carecovid-19-patients.

[34] Yang, G. Z., Nelson, B. J., Murphy, R. R., Choset, H., Christensen, H., Collins, S. H., ... & Kragic, D. (2020). Combating COVID-19—The role of robotics in managing public health and infectious diseases.

[35] Vaishnavi, P., Agnishwar, J., Padmanathan, K., Umashankar, S., Preethika, T., Annapoorani, S., & Subash, M. Artificial Intelligence and Drones to combat COVID-19.

[36] Miner, A. S., Laranjo, L., & Kocaballi, A. B. (2020). Chatbots in the fight against the COVID-19 pandemic. npj Digital Medicine, 3(1), 1–4.

[37] Oniani, D., & Wang, Y. (2020). A Qualitative Evaluation of Language Models on Automatic Question-Answering for COVID-19. arXiv preprint arXiv:2006.10964.

[38] Maniou, T. A., & Veglis, A. (2020). Employing a chatbot for news dissemination during crisis: Design, implementation and evaluation. Future Internet, 12(7), 109.

[39] Bharti, U., Bajaj, D., Batra, H., Lalit, S., Lalit, S., & Gangwani, A. (2020, June). Medbot: Conversational Artificial Intelligence Powered Chatbot for Delivering Tele-Health after COVID-19. In 2020 5th International Conference on Communication and Electronics Systems (ICCES) (pp. 870–875). IEEE.

[40] Martin, A., Nateqi, J., Gruarin, S., Munsch, N., Abdarahmane, I., & Knapp, B. (2020). An artificial intelligence-based first-line defence against COVID-19: digitally screening citizens for risks via a chatbot. bioRxiv.

4

Technological Opportunities to Fight COVID-19 for Indian Scenario

Meera C. S.[1], Aslesha Bodavula[2], and Pinisetti Swami Sairam[3]

[1]Departmnet of Electrical and Electronics Engineering, University of Petroleum and Energy Studies, Dehradun, Uttarakhand, India
[2]Kings Cornerstone International College, Chennai, Tamil Nadu, India
[3]School of Business, Woxsen University, Hyderabad, Telangana, India
Corresponding author: Pinisetti Swami Sairam, swami.sairam@gmail.com.

Abstract

The COVID-19 outbreak has created a surge of infections around the world. With a high transmission rate, the virus has created a major health concern especially for the people who are involved for its diagnosis and testing. The safety of health and frontline workers are of great significance during any pandemic situation. This article gives an insight of role of technology initiatives that could be actively implemented to reduce the impact of pandemic situations in the Indian scenario. These technologies can help to contain the devastating effects of the outbreak in faster and efficient ways.

4.1 Introduction

COVID-19 is a positive single-stranded ribonucleic acid (RNA) virus, which is a type of coronavirus that damages the respiratory system. Initially, the outbreak was detected in Wuhan, Hubei Province, China in December 2019 with the major symptoms of fever and dry cough [1]. The transmission of the virus occurs from person to person, which means people are in very close contact or through respiratory droplets formed at the time an infected

45

human being coughs or sneezes [2]. India is the second largest populated country with over 1.3 billion population and faces unique challenges to fight COVID-19. The first case in India was reported on 30 January 2020 who had travel histories to countries like China, Italy, and Iran. As the primary effort to flatten the pandemic curve, the country has announced a nationwide lockdown for 21 days. The huge density of population makes the risk of disease spread high especially in second- and third-tier cities. With the lack of medical professionals and modern infrastructure, the country could use more of technological solutions extensively to fight the infection. As of August 2020, the present status of infection in the country is shown in Figure 4.1 with more than 20 lakh confirmed cases [3].

As the pathogen is from the family of coronavirus [1], many countries are implementing the strategies developed during the Severe Acute Respiratory Syndrome Coronavirus (SARS-CoV) outbreak, such as protecting the staff with appropriate personal protective equipment (PPE), prevention of virus transmission, and reengineering the workflow to minimize exposure time for cross-infections. However, the high workload with an increasing number of cases and lack of PPEs and facilities such as isolation wards with negative air pressure and good ventilation systems may adversely affect the workers.

In the Indian scenario, it is evident that following timely disinfection and maintaining less contact time with the patients are very critical to reduce a catastrophe. The global transmission of novel coronavirus (SARS-CoV-2/COVID-19) has triggered the usage of the technologies effectively to mitigate

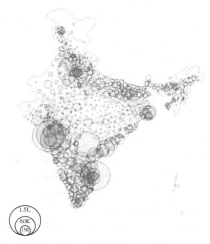

Figure 4.1   Infection status in India, April 2020 [3].

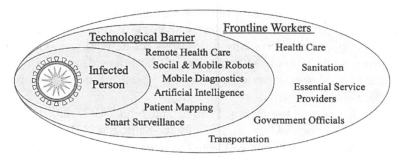

Figure 4.2 Applications of emerging technologies in managing the pandemic.

the effect of COVID-19 pandemic. With escalating mortality rates, adopting technological solutions could be of significant interest in the country [4]. Figure 4.2 shows the applications of technologies in managing the pandemic.

4.2 Technological Interventions

In many countries, the application of technologies like AI has helped in the contact tracing, tracking, and preparedness for COVID-19 so as to curb the spread of disease. Social media, machine learning (ML) models, and mobile phones were used as effective tools in preparing migration maps, location tracing, and movement tracking. ML models developed were used to forecast the transmission and enable surveillance [5]. The rising heart rates were captured using smart watches to identify the outbreaks. Also applications embedded in these smart watches were used to collect pulse and sleep pattern in addition to temperature to monitor the disease [6, 7]. With the tech interventions in healthcare, a low mortality rate could be maintained. In this article, we outlay the effective technologies that empower the healthcare systems and make a difference.

4.2.1 Robotic Technologies in COVID-19

The widespread pandemic could bring a substantial paradigm shift with healthcare robots to minimize the risks of human infection. Although few countries are already exploring the application with robots in healthcare, it is however limited. The medical and other healthcare workers are most vulnerable owing to the proximity with the infected patients. Physical examination is still a challenge in many places and PPE kits are cumbersome. Robotic arms can be equipped with microphones, cameras, and stethoscope and used in some places. They help to acquire vital information from patients kept in isolated areas avoiding direct contact with medical professionals and reduce the risk of infection. The technology of robot navigation could be used in applications pertaining to logistics for delivering food items and transporting drugs around. As robots cannot be infected, using a robot for collecting clinical specimens could not only speed up the process but also reduce the risk of exposure.

Robots like RoboDoc, Mitra, Prithvi, and NIGA-BOT are designed to reduce the risk of doctors from contracting the pathogen [8–10]. They help to remotely monitor the patients and scan the temperature and the other vitals. This helps to reduce the risk of infection occurring through contact. The humanoid robot ELF tested at AIIMS Delhi, one of the premier institutes in the country, was 92-cm-tall and can travel independently at 2.9 kmph, interact with the patients remotely, and bring down the risks of infection. Robots used in such purposes are shown in Figure 4.3. It enables patient interaction with doctors and even relatives from time to time using a 10.1" display-screen. It is embedded with more than 60 sensors and has an auto charging capability. Another humanoid robot, ITI Cuttack CO-BOT also ensures the protection of health workers. It has a hand-like structure and can move around the hospital carrying essential supplies up to 20 kg.

Also, requirements to reduce virus transmission through high touch-surface in hospitals have led to the development of sterilization robots emitting UV rays. Robots like Milagrow iMap 9 and Invento C-Astra are used to disinfect the ward floor of hospitals. With the high rate of infection, the technology is slowly spread to various hospitals across the country. Majority of the sanitation robots are having the autonomous motion feature enabled through Light Detection and Ranging (LIDAR) and Simultaneous localization and mapping (SLAM) technology. Also, they can perform cleaning of zones sequentially and virtual blocking of objects. Irradiation

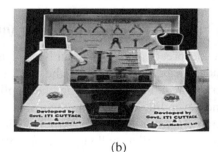

(a)   (b)

Figure 4.3   Robots developed to assist healthcare workers. (a) Humanoid ELF used by AIIMS Delhi. (b) CO-BOT developed by ITI Cuttak.

through UVC and sometimes even sodium hydrochloride solution are the features provided by these sanitation robots.

Another significant field that used robots is to raise awareness among the people about the spread of the virus. Asimov humanoid robots were used to brief the people about the preventive measures and the importance of social distancing and also to distribute sanitizers and masks [11]. Similar to thermal scanner-based systems that are used to monitor the temperature of multiple individuals in public spaces, mobile robots can be used to check the in/out-patient temperature. The robots are equipped with thermal scanners (or temperature guns) and ensure minimum contact with people.

4.2.2 Smart Surveillance Systems

To curb the spread of virus, one of the important practices adopted by the offices and other establishments is making thermal scanning a mandatory procedure for entering and exiting. The thermal cameras used to detect the high body temperature are split into thermography and surveillance module. Thermal screening uses the principle of the radiation from the body. A rise in body temperature increases the emitted amount of radiation from an object. Therefore, with thermography, one can see variations in temperature. Combined with visual cameras, the thermography systems are widely used in airports, public areas, hospitals, and other establishments for surveillance and fever monitoring.

Smart infrared thermal scanners are installed at the entrances of the establishments and institutions which indicate if the person has elevated body temperature, a symptom of COVID-19, as shown in Figure 4.4.

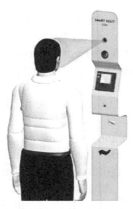

Figure 4.4 A smart assist kiosk monitoring mask wearing, thermal screening, and automatic sanitizer dispensing [12].

The systems automatically detect high body temperature of people (above average temperature) when the field of view of camera covers the face of the person. Generally, the temperature of forehead, retina, or inner canthus is measured as these regions give the temperature closer to core body temperature. If diagnosed with high temperature greater than threshold (approximately greater than or equal to 99.50 F), the detected personals are selected for additional screening and other specific tests for virus. The mini electronic modules do not need human presence or intervention to perform scanning and can be fixed to entry points of buildings, schools, and other public and commercial spaces. Various institutions have set up kiosks with advanced features of pulse, respiration rate, and blood oxygen screening in addition to temperature scans. They also incorporate features like face mask detection, automatic hand sanitization, and attendance management. Also, some of these modules provide cloud storage and QR scanning as add-ons.

The recent researches are conducted to explore thermal imagers in smartphones and even smart watches. A temperature measurement feature in a smartphone would help people to measure the body temperature frequently without the need of carrying an extra device. The size and reliability of these thermal imagers and detectors are significant areas of research in these electronic modules. The size should be small enough to be able to integrate into mobile and an accuracy of not greater than 0.5°C is necessary for a reliable operation. Also, ways to reduce false alarms, cost effectiveness, and educating the consumers on interpretation/reading of the thermal images are also matters of study.

4.2.3 Artificial Intelligence and Machine Learning

Application of ML algorithms has been significant in predicting the risk of disease in terms of infection, severity, and outcome risk. Infection risks deal with the probability that a particular person or group getting infected, severity risk is the probability that a particular person or individual requires intensive care or hospitalization after getting infected, and outcome risk is the probability that a particular treatment will be ineffective or will lead to death for an infected person.

Artificial intelligence (AI) and ML have gained attention for its possible applications toward controlling the pandemic via early detection and monitoring, contact tracing of individuals, and development of drugs and vaccines. A study conducted on AI applications' use in this pandemic showed that more than 35 countries have built applications particularly for contact tracing which is an essential method to break the chain and has become an

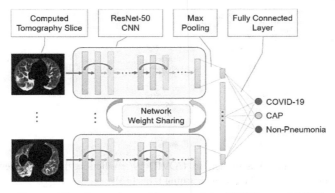

Figure 4.5 Illustrative architecture of the COVNet model [13].

important step in prevention of the disease when a person is diagnosed and confirmed for COVID-19. Algorithms such as support vector regression and stacking-ensemble, deep learning using Long short-term memory (LSTM) network, and regression tress were being used to generate early short-term forecasts of COVID-19. These forecasts helped the healthcare experts, policymakers, and the governments to prepare themselves in combating the virus. Three kinds of CT images, including COVID-19, community acquired pneumonia, and other non-pneumonia cases, are mixed to test the robustness of the proposed model, which is illustrated in Figure 4.5.

Although standard methods provide first results of any tests being conducted, these often require considerable amount of time and resources. Use of ML techniques can help in reducing the necessary time frames producing automated analysis and allowing AI researcher to further support and upgrade the systems. A recent supervised classifier trained on a large dataset of chest X-rays achieved a mean area under the receiver operating characteristic curve of 94% for the diagnosis of 14 distinct lung pathologies [14]. Furthermore, the preliminary studies based on a few hundred chest CT scans suggest that COVID-19 can be automatically diagnosed with ML. Figure 4.6 shows the incorporation of natural language processing (NLP), ML, and epidemic SIR (Susceptible, Infected and Recovered) model integration for better prediction.

However, the use of ML of medical images to diagnose or prognose COVID-19 remains currently limited. With rising cases worldwide and every hospital and community centers being converted to treat the infected peoples, challenges to address and treat the other diseases have been a task for the healthcare workers. Use of AI and ML in developing applications such as chatbots, telepresence robots have increased to cater the needs.

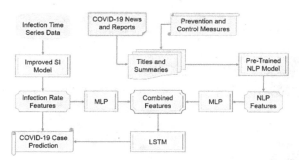

Figure 4.6   An AI-based approach to COVID-19 prediction that combines traditional epidemic SI model, NLP, and machine learning tools [13].

4.2.4 Computational Fluid Dynamics

Forecasting contagion risk due to human-produced airborne pathogens in healthcare amenities needs a comprehensive knowledge of droplet generation and dispersion mechanisms. Computational fluid dynamics (CFD) analysis can play a significant role to understand the effect of the droplet dispersion and transmission of the pathogen virus due to sneezing and coughing in the isolation wards and the other regions in the hospital. Numerical analysis carried out to check the impact of the sneeze with no protection, with an elbow, and with a mask using the Cradle scFLOW, in a room with two persons standing with a distance of 2 m can be seen in Figure 4.7. From the analysis, it was concluded that with an elbow, the droplets reached half of the distance and try to settle down on the ground. While with the mask scenario, the droplets did not reach 1/4th of the distance [15].

A CFD study was carried out in the metro/train car to duplicate the physical behavior of the airborne droplets, when the breathing and speaking persons close to each other with a distance apart approximately 0.7 m using Hexagon & Cradle. Two scenarios are considered to perform the simulations, one is a speaking person without a mask and another one is with a mask. From the results, it was found that the speaking person with a mask has the potential to reduce airborne droplets that are blown out to the environment and the droplets getting close to the opposing person were significantly degraded [16].

During a pandemic, the major concern is about the hospitals and the environment in the hospital surroundings. In these situations, the patients are treated in the isolation wards to break the transmission of the virus to the other patients present in the hospital [17]. Numerical analysis of hospitalization of the first Middle East respiratory syndrome (MERS) patient was carried out to

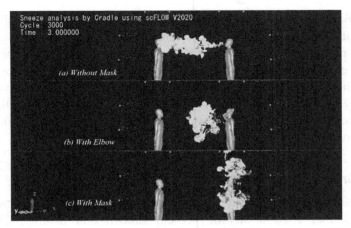

Figure 4.7 Sneeze analysis by Cradle using scFLOW. (a) Without mask. (b) With elbow. (c) With mask [15].

study the transmission efficiency of the virus through airflow and the effect of ventilation system. From the numerical analysis, it was concluded that virus transmission is probable through airflow in poor ventilation systems [18]. Providing the proper ventilation system in these rooms can have the potential to slow down the spread of the virus through airflow and can also assure a better health condition of the patients. Study of droplet contact with ventilation system flow-pattern was carried out to extract the droplets through exhaust and reduce the risk of transmission of airborne infectious diseases. Mainly in ventilation design for healthcare facilities, this mechanism can diminish airborne infection risk. The numerical analysis of isolation rooms could provide the airflow circulation inside the rooms and can lead to better design of the ventilation system in the hospitals.

Conducting the thermal analysis can help us to know about the comfort of the patients in the operating rooms and isolation wards [19]. This could slow down the spread of the virus over airflow and can also assure better health conditions in hospitals. Mechanical ventilation systems play a major role in breathing comfort of patients with respiratory illness. The numerical analysis using the boundary conditions of the patient could help to understand the required flow rate, rise, and reduce the time of the pressure inhalation and exhalation. This analysis can help to guide clinicians to operate the ventilation systems in pandemic situations.

With the help of the CFD with real-time scenarios, we can study the airflow analysis in closed places, hospitals, office buildings, at shopping malls, airports, and other crowded locations. By using CFD, engineers are accomplished to study the air movements in locked places by excluding air recirculation regions, saving energy through added effective heating and/or cooling, reducing the dispersion of bacteria by getting rid of airborne bacteria through effective ventilation, improving thermal comfort levels for patients and medical staff, quickly regulating temperature without large overshoots. Using CFD, we can study the physical behavior of the person with a mask and without a mask at various types of places like hospital, home, metro/train car, airports, etc.

4.2.5 Unmanned Aerial Vehicles

Unmanned vehicles like drones have emerged as a key tool in fighting against the pandemic. Their ability of Vertical Takeoff and Landing Capability (VTOL) and Beyond Visual Line of Sight (BVLOS) are highly beneficial at the time of crisis. Also, the portability and, in emergencies, their ability to takeoff and provide assessment of the condition, including visual data and extent of damage, are ideal in such critical situations. They can be remotely piloted and are therefore effective in minimizing interactions which comes as a boon when the authorities meant to safeguard the communities and ensure regulations are becoming highly susceptible to contacting the disease. The main areas where the drone technology plays an effective role in combating the virus spread are surveillance and broadcast, delivery of supplies, disinfection, and temperature check.

As an important measure, the authorities across the world has banned large gatherings as well as closed down public places that are non-essential. These steps were adopted to ensure social distancing and to reduce people-to-people contact. However, in places where the people are not adhering to the restrictions either knowingly and unknowingly, the law enforcement authorities are depending on drones for street monitoring as well as breaking social gathering that could increase the chance of virus spread. These actions are done without any physical engagement and thus benefit the frontline workers from the risk of getting infected. Furthermore, drones are proven to be effective in broadcasting messages regarding lockdown measures especially in remote and rural areas where there is no open communication channel for health information. In countries like India, China, and several European countries, drones equipped with large megaphones and a cell phone are used to make broadcasting and announcements to public.

With recent findings that the virus may survive on hard surfaces from few hours to several hours, the need to disinfect public spaces is high. In several places, the authorities are using the technology of spraying drones where

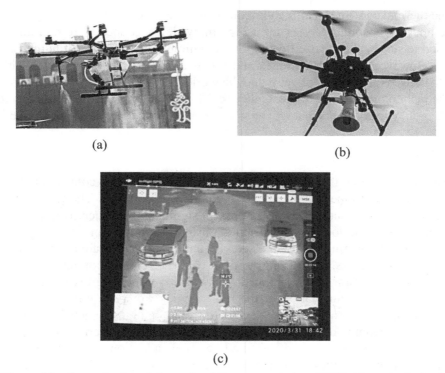

(a)

(b)

(c)

Figure 4.8 Customization of drones for different uses during COVID-19 pandemic for (a) disinfection [20], (b) announcement [21], and (c) thermal scanning [22].

they are filled with disinfectants and are used to spray to the public areas as shown in Figure 4.8. The benefits of using spraying drones are the speed of operation and the area covered compared to conventional methods. These drones are inexpensive, can be mobilized swiftly, and are easy to operate.

Though not implemented on full scale, one of the fastest and safest ways to deliver essential medical supplies is through the use of drone technology. The drones are Internet of Things (IoT) enabled and structure drones could be easily modified to accommodate a payload dropping mechanism to deliver the supplies. They reduce transit time and enhance contactless handovers, thereby reducing the risk of infection. Other than the medical supplies, groceries are also being delivered through drones. These are particularly useful for delivering in hotspots where there is a high risk of infection.

Drones with infrared cameras are remotely used to carry out large-scale temperature checking. This avoids face-to-face contact and reduces the risk of infection, especially where lockdown was implemented, while calibration of the thermal camera remains an important feature. Also in the countries where temporary hospitals are set up using empty fields, drones were constantly used to survey the areas efficiently with minimum involvement of humans.

However, the use of drones and other unmanned vehicles comes with challenges on rights of privacy, and the world has begun to see the benefits of the technology. In case of emergency services, there is an active call to revisit the policy frameworks that will enable to help to provide emergency services.

4.3 Conclusion

For an infection that keeps everyone in direct/indirect contact at risk, technology has emerged as a hope to combat the outbreak. The technology cannot stop the pandemics spread but can significantly reduce its impact on society. With the lockdown restriction easing, these technologies will play key roles in ensuring safety and security and also to facilitate arrangement, building awareness, monitoring, testing, contact locating, confinement, and clinical supervision have continued front-runners in dealing disease burden. Though the implementation of the technologies would require the several policy loops to ensure data privacy, there would be high impacts in the long term. The benefits of the listed technological solutions in terms of cost and comfort need to be assessed before a large-scale adoption. We anticipate the implementation of these initiatives to contain the pandemics.

References

[1] C. Huang *et al.*, "Clinical features of patients infected with 2019 novel coronavirus in Wuhan, China," *Lancet*, vol. 395, no. 10223, pp. 497–506, Feb. 2020, doi: 10.1016/S0140-6736(20)30183-5.

[2] J. Wei and Y. Li, "Human cough as a two-stage jet and its role in particle transport," *PLoS One*, vol. 12, no. 1, Jan. 2017, doi: 10.1371/journal.pone.0169235.

[3] "Coronavirus Outbreak in India - covid19india.org." https://www.covid19india.org/ (accessed Aug. 13, 2020).

[4] B. Krishnakumar and S. Rana, "COVID 19 in INDIA: Strategies to combat from combination threat of life and livelihood," *J. Microbiol. Immunol. Infect.*, vol. 53, no. 3, pp. 389–391, Jun. 2020, doi: 10.1016/j.jmii.2020.03.024.

[5] The Washington Post, "How digital data collection can help track Covid-19 cases in real time." .

[6] T. Reuters., "Germany launches new smartwatch application to monitor coronavirus spread." .

[7] S. Whitelaw, M. A. Mamas, E. Topol, and H. G. C. Van Spall, "Applications of digital technology in COVID-19 pandemic planning and response," *The Lancet Digital Health*, vol. 2, no. 8. Elsevier Ltd, pp. e435–e440, Aug. 01, 2020, doi: 10.1016/S2589-7500(20)30142-4.

[8] "Mitra - Invento Robotics." https://mitrarobot.com/ (accessed Aug. 13, 2020).

[9] "ITI Cuttack develops low-cost robots to combat COVID-19 - The Hindu." https://www.thehindu.com/news/national/other-states/iti-cuttack-develops-low-cost-robots-to-combat-covid-19/article31463832.ece (accessed Aug. 13, 2020).

[10] "Milagrow HumanTech." https://milagrowhumantech.com/ (accessed Aug. 13, 2020).

[11] "Asimov Robotics." https://www.asimovrobotics.com/ (accessed Aug. 13, 2020).

[12] "'Smart Assist' One Stop Kiosk for COVID19 Precautions." https://www.themachinemaker.com/innovation/ab-plastomech-avadhoot-automation-innovates-smart-assist-covid19 (accessed Aug. 13, 2020).

[13] T. T. Nguyen, "Artificial Intelligence in the Battle against Coronavirus (COVID-19): A Survey and Future Research Directions," doi: 10.13140/RG.2.2.36491.23846.

[14] "Computational predictions of protein structures associated with COVID-19. Deepmind," 2020. .

[15] "Work from Home - Stay Safe." https://www.mscsoftware.com/work-from-home/stay-safe (accessed Aug. 13, 2020).

[16] H. & C. Softwares, "CFD Analysis of Metro Car (COVID-19)," 2020. https://www.mscsoftware.com/sites/default/files/CFD-Metro.pdf (accessed Aug. 13, 2020).

[17] S. K. M. Rao, "Designing hospital for better infection control: An experience," *Med. J. Armed Forces India*, vol. 60, no. 1, pp. 63–66, 2004, doi: 10.1016/S0377-1237(04)80163-1.

[18] S. Jo, J. K. Hong, S. E. Lee, M. Ki, B. Y. Choi, and M. Sung, "Airflow analysis of Pyeongtaek St Mary's Hospital during hospitalization of the first Middle East respiratory syndrome patient in Korea," *R. Soc. Open Sci.*, vol. 6, no. 3, Mar. 2019, doi: 10.1098/rsos.181164.

[19] F. Memarzadeh and P. E. A. Manning, "Thermal Comfort, Uniformity, and Ventilation Effectiveness in Patient Rooms: Performance Assessment Using Ventilation Indices."

[20] "Drones and the Coronavirus: Do These Applications Make Sense? (Updated) - WeRobotics Blog." https://blog.werobotics.org/2020/04/09/drones-coronavirus-no-sense/ (accessed Aug. 13, 2020).

[21] "Africa is using technology to innovate solutions to COVID-19 | World Economic Forum." https://www.weforum.org/agenda/2020/07/african-innovators-coronavirus-drones-phones/ (accessed Aug. 13, 2020).

[22] "Cyberabad Police on Twitter: "Cyberabad police is taking all possible measures to enforce the state-wide #lockdown & #stopthespread of #Coronavirus. We are using @Cyient's drone-based inspection capability to monitor the situation through drone cameras & thermal imaging technology". #essentialservices https://t.co/VuVSeAhAfs" / Twitter." https://twitter.com/cyberabadpolice/status/1245598701860446210 (accessed Aug. 13, 2020).

5

Mobile Robots in COVID-19

Prashant Kumar Dwivedi

The Hi-tech Robotic Systemz Ltd, Gurugram, Haryana 122001, India
Email: dwivedi.kr.prashant@gmail.com

Abstract

Robotics as a discipline focuses on combining perception with actions for human assistance in day-to-day lives and keeping them safe. Before COVID-19, mobile robots were limited mostly to the manufacturing industry, but with the present scenario, both demand and necessity have increased. Social distancing was required at all levels to flatten the pandemic curve, but it also had its challenges for essential services. With fewer working hands, the necessity to maintain social distancing directed us toward innovations in mobile robotics. The installation minimized human contact, substituted human in cleaning, sanitization, and delivery, and also monitored interactions and meetings. Various applications were developed as POC (Proof of Concept)in these difficult times to ensure safety and distancing. Ultraviolet (UV) robots disinfected the spaces, sanitizer spray robots are used in the manufacturing area, food and medicine delivery robots for hospitals, and, similarly, interaction and follow-me robots interacted with patients. It minimized the impact of COVID in hospitals, quarantine facilities, and manufacturing area. The distance was maintained from contaminated zones and allowed the workforce to focus on important tasks and continue the fight. Some applications also utilized perception and deep learning technologies to identify people with masks and perform thermal screenings. Mobile robots' potential has been realized and the horizon has widened. It has thus created an opportunity for advanced research in healthcare mobile robotics and low-cost indoor robotics by using innovative technologies.

5.1 Introduction

Over the last decade, understanding of robotics and its application has transformed a lot. Starting with basic manipulator to mobile robots and autonomous cars, this field has seen a lot of advancement and research. Robotics as a practical application has been mostly limited to industrial uses and setup of robotic arms, while there is continuous research in the field of mobile robots and autonomous cars. The major disadvantage with the industrial robotics setup is mobility and ability to work in the shared space with humans. Hence, the requirement of mobile robots came into existence. With difficult situations such as COVID-19, different innovations have been made to make mobile robots more useful.

COVID-19 is an infectious disease caused by a newly discovered SARS-CoV-2, which is a novel coronavirus. On 31 December 2019, the World Health Organization (WHO) was formally notified about a cluster of cases of pneumonia in Wuhan City, which is home to 11 million people and is the cultural and economic hub of central China. The COVID-19 pandemic has spread with alarming speed, infecting millions and bringing economic activities to a near stand-still as countries imposed tight restrictions on movement to halt the virus. Along with economic activities, the pandemic caused significant disruption to daily life and social activities. At present, the best way to prevent contracting COVID-19 is to avoid being exposed to the coronavirus. Organizations such as the Centers for Disease Control and Prevention (CDC) [1] recommended guidelines including social distancing, wearing masks, and washing hands frequently to reduce the chances of contracting or spreading the virus.

The current scenario pointed out that the role of mobile robots is very crucial and essential in healthcare and societal sector. The present pandemic created a huge catastrophic effect throughout the world. This created an opportunity for developing new technology in the field of robotics and Internet of Things (IoT) technology for working alongside corona warriors. It is also time for new innovations in methodologies, concepts, and ideas to be developed further to prepare the world and stand in front line to serve the mankind [2].

5.1.1 What is Mobile Robot?

A mobile robot [3] is a machine controlled by set of software commands by using combination of sensors to understand about its surroundings and actuators to move around its environment.

Mobile robots are majorly called autonomous mobile robots because they are expected to move in the environment without any external help and perform their task.

To achieve mobility and planning abilities, mobile robot basics are divided into field of locomotion, kinematics, perception, SLAM (Simultaneous localization and mapping), and navigation. Locomotion and kinematics are important to move on different kinds of surfaces and handle different kinds of models. Perception deals with the understanding of area and surroundings and helps take decision based on sensors and computer vision technologies.

SLAM is simultaneous mapping and localization, i.e., mapping the area to get the information of places where robot has to run while localization is the field that deals with pose estimation in the given map. Once we have the pose, navigation helps us to take decision and run the robot based on a given task. Task can be anything based on application starting from industrial purposes such as material movement, production robots, co-bots, etc., to household application such as cleaning robot. Mobile robots also are taken into use for exploration purpose for places where it is not possible for humans to reach like mars rover, tunnel exploration robots, or underwater robots.

5.1.2 Components of Mobile Robots
5.1.2.1 Locomotion and Kinematics

Most of the mobile robots work in defined and controlled environment but not fixed to static environment. Sometimes, they have to maneuver through rough terrains and unknown environment leading to many ways to move in the environment, making locomotion an important aspect of mobile robot design. Locomotion selection depends on many things like stability, structure of environment, dynamics, path size and shape, center of gravity, friction, angle of contact, and geometry of contact points. Based on the requirement and challenges, mobile robots majorly use wheeled or legged mechanism.

Wheeled mechanism is the widely used mechanism as compared to legged or other methods as it is easier to design and integration complexity is very minimal. Wheeled mechanism is subdivided into the following four types:

1. *Differential drive*: Differential drive is the mechanism where two wheels are mounted on the common axis and each wheel can be controlled separately to move it forward/backward.
2. *Tricycle drive*: Tricycle, as its name suggests, consists of three wheels in which two castors are mounted on the common axis while the third wheel is steerable and powered to provide movement.

3. *Ackermann drive*: Ackermann is a car-like mechanism with four wheels where front wheels are steering type to move the wheels in and out.
4. *Omni wheel drive*: Omni wheel drive is holonomic in nature which means it can drive in any direction without changing the orientation of the robot.

The most commonly used mechanism is differential drive. Odometry [4] calculation is an important aspect that varies based on mechanism type and calculated based on kinematics of the mechanism, but there is calibration required to get proper value. Also, visual odometry [5] techniques are available to augment and, in some cases, replace odometry [6].

5.1.2.2 Perception
Acquiring the data of the given environment and extracting the useful information to understand the environment is an important task for autonomous mobile robots. Using multiple types of sensors like Camera, Lidar, Radar, IMU, Encoders, etc., sensor fusion [7] helps in reducing error and measurement accuracy of the algorithm. Perception is used to adapt changes in the environment and help in deciding the set of operations to achieve the goal. Computer vision and deep-learning-based methods are widely used in this field for object detection, scene identification, segmentation, face recognition, temperature detection, monitoring, and many applications in navigation [8] such as follow-me robot [9].

There are many kinds of sensors [10] that can be used to collect the data which are grouped into:

1. *Proprioceptive*: For monitoring internals of mobile robot-like battery, speed, etc.
2. *Exteroceptive*: For collecting data from robot environment like cameras.
3. *Passive*: Sensors which measure environmental energy entering the sensor like vision-based sensors, gyroscope, etc.
4. *Active*: Sensors which emit energy to environment like ultrasonic, reflective beacons, etc.

5.1.2.3 Slam
SLAM [11] helps to solve the "Where am I?" problem in the field of robotics. In order to work without any external help, it is important for a mobile robot to have a map of the place where it has to run along with the ability to localize itself in that map.

Mapping is a process of collecting the data from the sensor and converting it into an image or database to represent the environment as it looks in real world. Accuracy and precision or robot movement depend on quality of mapping process.

Localization is the process of knowing the exact position of the mobile robot in the mapped area and the data provided by the sensor is matched with the map or database to find the current position called pose. There are many SLAM algorithms [12] available, such as particle filter-based SLAM [13], cartographer SLAM, RTAB map SLAM [14], etc. Very few mobile robots use 2D sensors while others use camera and 3D Lidar to enhance the SLAM algorithms. SLAM algorithms are still struggling with issues like kidnapped robot, making it difficult for mobile robot to run in highly dynamic environment.

5.1.2.4 Path Planning and Navigation

Path planning focuses on providing an optimal collision-free path for the mobile robot to reach the final destination.

Path planning can be subdivided into three types:

1. *Global path planning:* Global path planning is used to find the optimized shortest path between the initial and the goal position. Global path does not consider dynamic obstacles and plan a path based on the map and mobile robot footprint. Shortest path algorithms like A* Dijkstra work as global path planner. Also, dynamic path planning [15] based on genetic algorithms and different algorithms based on Voronoi diagram [16] also can be used for global path planning approach.
2. *Local path planning:* Path provided by global path is divided into multiple points and to follow that path, local planners are used. Local path planner considers dynamic obstacles and motion preemptive to find the best motion parameters for following the global path, and, accordingly, using the kinematics equations, motor velocities are calculated and given to motor to follow the optimal path. Methods like DWA [17] (Dynamic Window Approach) and Trajectory Generation [18] are examples of local planner.
3. *Obstacle avoidance:* Obstacle avoidance is considered as a problem to avoid dynamic obstacle in the global path and generating a route around it or to decide if the obstacle is moving and the mobile robot shouts "wait" to let the obstacle pass.

5.1.3 Mobile Robots and COVID-19

With the combination of perception, SLAM, and navigation, we can prepare a mobile robot to run in a defined area but most of the research is academic-based as not many application domains were explored till now.

WHO declared COVID-19 pandemic as global emergency on 30 January 2020 and the world had no idea how to deal with the situation. It caused a significant disruption to daily life and caused significant economic and social impacts. Due to globalization and interconnected operations between countries, the whole world got affected with COVID-19. Healthcare industry [19] is fighting in the best possible way, but lack of infrastructure is an issue. With the guidelines issued by healthcare departments, few things to consider to avoid the transfer of the virus were social distancing which means a minimum safe distance has to be maintained to avoid virus transfer from one body to other, no physical contacts, avoiding surface contacts without sanitization. After all the care and precautions also, it was very difficult for various professions and front-line fighters to avoid direct contact and maintaining sanitization at frequent intervals.

With this, the following question arises: Could mobile robot be a solution in combating COVID-19 [20]? Mobile robots have potential to do many tasks but never explored in the healthcare industry at this large scale. Mobile robots are capable of disinfecting, delivering medicines to patients, remote monitoring [1], and consultancy between doctor and patient and vitals monitoring at places like offices, airports, and other public places.

As robotics is a developing field, many innovations and researches are done during these times and many mobile robots, drones, and AI-based robots are deployed to fight in the front line with healthcare people [2].

At the same time, many applications have been developed to suit the industries for disinfecting the manufacturing line and process areas to get back the industries back to track with proper safety.

5.2 Requirements of Mobile Robots in Pandemic Situation

The applications of mobile robots in healthcare [20] and other allied areas are experiencing growth day by day. The International Federation of Robots (IFR) predicts an ever-increasing trend in the demand of medical robots within the coming years with an estimation of 9.1 billion USD market by 2022. Mobile robots not only assisted essential workers but were also used to carry out complex and precise tasks to reduce workload while maintaining accuracy and safety.

1. **Operator/staff safety:**
   1.1. Mobile robot works in collaboration with medical staff and operators, enabling medical staff to avoid direct patient contact.
   1.2. Mobile robot itself should have all the safety features to avoid any kind of accident. It should be safe enough so that staff should feel confident working alongside the robots in the same premises.

2. **Social distancing/disinfecting:**
   2.1. In order to prevent the spread of the virus, mobile robots prevent direct contact with patients in hospitals by delivering food, medicine, and other essential items.
   2.2. At public places, mobile robots with mask detection features and social distancing feature to detect and generate warning for social distancing.
   2.3. In manufacturing process, replacing humans for disinfecting the area.
   2.4. By attaching an ultraviolet (UV) light or defogger, mobile robot should be able to maneuver around the premises to kill viruses and disinfect the surfaces.
3. **Flexibility of time/operation:**
   3.1. Time-based mission and task allocation is one thing which can be achieved easily with mobile robots without any human help. Call-based features can also be provided for delivering the material and disinfecting the area.
4. **Ease of handling:**
   4.1. In most cases, robots in hospital and process areas have to be operated by non-technical staff, mostly medical staff and production supervisor; in that case, architecture and standard operating procedure has to be easy.
   4.2. Operation of mobile robot should be a single instruction or click based to avoid complex inputs from the operator.
   4.3. Handling, maintenance, and basic troubleshooting should be easy and fast.
5. **Contact-less monitoring**
   5.1. In hospitals, mobile robot fitted with IoT device should be able to reach the commanded destination, i.e., patient bed location so that video iteration can happen between doctor and patient.
   5.2. Operator should be able to monitor and take vital information of the patient without coming in contact with the mobile robot.
   5.3. At places like airport, mobile robots should do vitals monitoring to avoid contact with the operator.

5.3 Innovation and Classification of Mobile Robots

Mobile robots in COVID are classified based on the applications which are widely distributed over heath industry, manufacturing process and public places like airport etc.

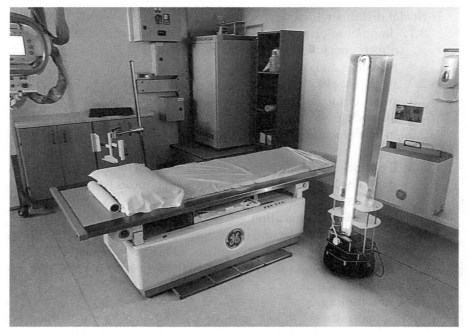

Figure 5.1   UV disinfection robot.

1. UV Robot: UV robot is used for disease prevention and robot-controlled non-contact UV surface disinfection as COVID spreads not only from person to person but also via close contact. The virus easily persists on metal, glass, and plastic for various days. UV light devices [21] reduce contamination on high touch surfaces. Instead of manual disinfection, which requires workforce mobilization and increases exposure risk to cleaning personnel, autonomous or remote-controlled disinfection robots could lead to cost-effective, fast, and effective disinfection. UVD robot [22] (UVD Robots ApS, Odense, Denmark) is an ultra-violet radiation-based device used to disinfect hospital. Akara robotics used research platform TurtleBot and converted into autonomous UV disinfecting robot.
2. **Telepresence Robot**: These robots teleport the user to a remote location using wireless Internet connectivity. It has features like face detection and follow-me to recognize and follow the user to improve face-to-face or screen-to-screen communication. For example, Beam Robot [23] helps in safe navigation in both spacious and confined environments with assisted docking feature. The robot can be controlled with both

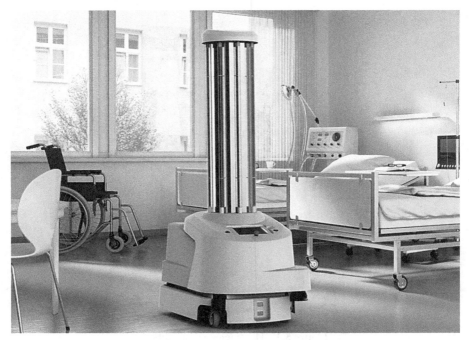

Figure 5.2 TurtleBot UV disinfectant robot.

phone and desktop application. Similarly, Temi [24] allows the user to add AIs such as Alexa to interact with humans providing with dynamic audio–visual experiences.

3. **Cleaning Robot**: Cleaning robots are used for dry vacuum and mopping. They form an integral part of healthcare industry to remove germs and pesticides. For example, Roomba [25] cleaning robot (iRobot, Bedford, MA, USA) is an intelligent navigating vacuum pump for dry/wet mopping. Peanut robot [26] (San Francisco, USA) uses highly dynamic robotic gripper and sensor system to clean washrooms of hospitals.

4. **Delivery Robot**: They are very useful in transporting/delivering essentials which range from medicines, food, and other items useful for inbound patients with coming in direct contact and maintaining social distancing.

   These robots are fully autonomous with man-in-the-loop operation where an operator can operate them at a distance. For example, Hi-tech Robotic Systemz provided delivery robots to healthcare facility [27] in COVID-19, limiting the human contact to deliver food and medicine. It kept the healthcare professionals safe from possible risk of infection.

Figure 5.3   Beam telepresence robots.

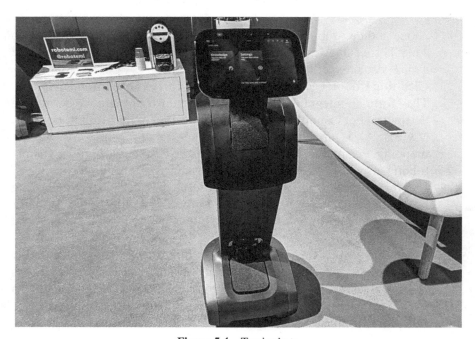

Figure 5.4   Temi robots.

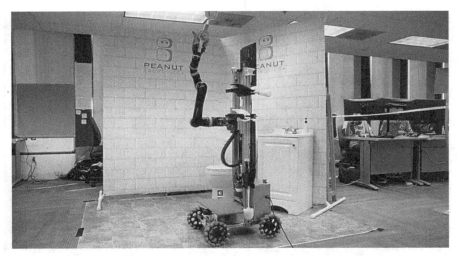

Figure 5.5   Peanut robot.

Figure 5.6   Roomba cleaning robot by iRobot.

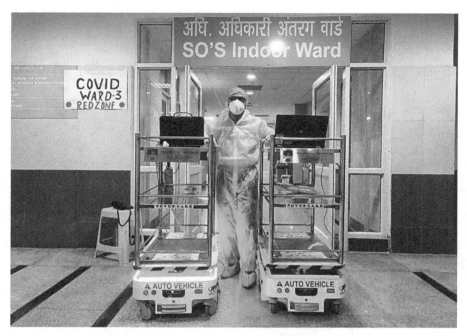

Figure 5.7 Delivery robots in healthcare facility.

Figure 5.8 Delivery robot.

Figure 5.9 Defogger robot by Hi-Tech Robotics.

5. **Spray/Defogger Robot**: These robots are used to spray disinfectants over large outdoor and indoor areas. They are remotely connected to avoid any hazardous contact with the spray such as residential areas. For example, Hi-tech Robotic Systemz installed spray robots in various production line which is time based to avoid any hazardous contact with workers. It includes features such as autonomous navigation, obstacle avoidance, mission selection, and time-based mission.

6. **Nurse Robot**: These robots assist medical staff in the healthcare in the same manner as that of human nurses.

   Nurse robots are commonly used in Japanese hospitals as Japan has the highest percentage of elderly (above 75 years) individuals among OECDcountries as nursing and healthcare individuals undergo high stress and exhaustion due to patient load. For example, Tommy is one of the six robots helping flesh and blood doctors and nurses care for coronavirus patients at the Circolo Hospital in Varese, a city in the northern Lombardy region that was the epicenter of outbreak in Italy.

5.4 Future Scope and Challenges

Mobile robots attend to the individuals who are non-compliant with social distancing norm. All individuals in an indoor environment are encouraged to maintain a 6-feet distance from one another. More studies are required on the social impacts of mobile robots. Due to COVID, mobile robots have performed in low to medium density settings. The future scope is to design better techniques to improve the enforcement of social distancing by using advanced human–robot interactions. COVID-19 is a catalyst for developing robotic systems that can be rapidly deployed with remote access by experts and essential service providers without the need of traveling to front lines. Widespread quarantine of patients may also mean prolonged isolation of individuals from social interaction, which may have a negative impact on mental health. To address this issue, social robots could be deployed to provide continued social interactions and adherence to treatment regimens without fear of spreading disease.

However, this is a challenging area of development because social interactions require building and maintaining complex models of people, including their knowledge, beliefs, emotions, as well as the context and environment of the interaction.

At present, the collaboration between human and robots is limited. The future scope in mobile robotics is more user friendly and collaborative robots with improved perception and navigation algorithms. The R&D team is continuously working toward developing mobile robots that overcome challenges in payload, navigation, and cost [28].

5.4.1 Challenges During Development Phase

Availability of infrastructure – manufacturing of robots is a costly process and it is still a part of research throughout the world. In terms of availability of resources, there are very few educational institute and industries with proper resources to do research and testing. A better infrastructure and education system in the field of robotics is required especially in developing countries like India.

Cost-effective solution – the design and installation should be cost-effective. Since the solutions have to be implemented on a larger scale, they must be cost-effective. High cost may make them unfeasible for startups and small- and medium-scale industries. Since the idea is to widely spread the use of mobile robots, it is one major challenge to work on cost-effectiveness.

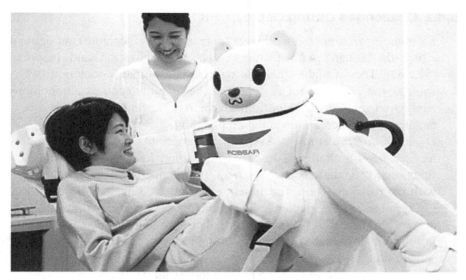

Figure 5.10 Robot carers in Japan.

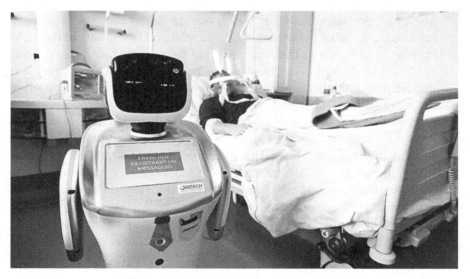

Figure 5.11 A robot helping medical team.

5.4.2 Challenges During Deployment Phase

Dynamic environment – mobile robots are installed in sociable environment alongside humans with features such as localization and mapping (SLAM). They are able to work with certain human interaction but as soon as the environment is dynamic, the features such as autonomous navigation get affected. Algorithms have to be designed for mobile robots enabling better perception and path planning.

Dynamic task allocation – Assigning appropriate tasks to mobile robots has not been a challenge so far. To a certain level, mobile robots are functional in situations such as loop, pre-planned which are integrated with the mission. These functions can be selected in the mission at any given point, but the challenge lies in the allocation of dynamic tasks. At present, in the case of dynamic task, there are human interventions, but in future, it is expected to reduce through intense research in the area of scheduling.

Slow operational speed – With present industrial standards, mobile robots are operated as speed way less than the speed of the human. Since mobile robots are sharing the workload, it has to take care of the operational speed. Assembly of a product often follows a sequence of actions called assembly precedence. When collaborating with another production resource (e.g., robot or human), the sequential processes must be balanced in time to take care of the preceding and subsequent tasks to avoid any bottle necks. If mobile robots have to be used as co-worker, then it is important to attain the same speed as humans. It requires anticipation of the movement and change paths in case of collision.

Deployment time – the deployment time is usually higher than anticipation especially in complex tasks such as assembly. The unpredicted situations in planning stage are related to planning, need of grippers, and other safety complications. The movements of the robots are coordinated with other hardware, peripherals, and humans since they do not perform in isolation. All of these increase integration and operational complexity.

Reprogramming and reconfiguring – one of the major issues faced after the deployment of the robots is reprogramming and reconfiguring the mobile robot in a facility without any technical person. These situations are mostly unaccounted for during the planning phase; thus, it becomes really difficult to do the entire process especially in situations like COVID. It is, thus, important to work on various interfaces to avoid such situations in industries.

5.5 Conclusion

The above chapter presents a comprehensive overview of mobile robots in healthcare sector and other areas with special relation to the assistance and control the spread of COVID-19 pandemic. As stated by WHO, COVID-19 can only reduce through effective management by reducing the number of infected patients. The situation has turned out to be a global challenge and mobile robots have provided great assistance being a helping hand to essential workers. Since the costing is still a big challenge, technologically advanced countries can contribute toward advanced infrastructure. There are other challenges that have been discussed above which are faced during deployment and development face; therefore, R&Ds of companies have to be funded to avoid such challenges in future and be prepared for future situations. Opportunities are also available in the design and operations of mobile robots using optimized algorithms, which also have to be explored in the future to benefit the society in an overall manner.

References

[1] J. A. Sathyamoorthy, U. Patel, A. Savle, M. Paul and D. Manocha, "COVID-Robot: Monitoring Social Distancing Constraints in Crowded Scenarios," *arXiv:2008.06585v2*, pp. 1–11, 2020.

[2] R. K, M. Gopikrishna, V. R. Rao, P. Pavani and C. Chandrasekhara, "Smart Applications using Robotic and Iot Technologies in Fighting against Pandemic Covid19 in Medical and Societal Sectors," *International Journal of Innovative Technology and Exploring Engineering (IJITEE)*, vol. 9, no. 7, pp. 2278–3075, May 2020.

[3] F. Rubio, F. Valero and C. Llopis-Albert, "A review of mobile robots: Concepts, methods, theoretical framework, and applications," *International Journal of Advanced Robotic Systems*, no. March-April, pp. 1–22, 2019.

[4] E. Papadopoulos and M. Misailidis, "On Differential Drive Robot Odometry with Application to Path Planning," in *Proceedings of the European Control Conference*, Kos, Greece, 2007.

[5] C. Forster, M. Pizzoli and D. Scaramuzza, "SVO: Fast Semi-Direct Monocular Visual Odometry," in *IEEE International Conference on Robotics and Animation*, Hongkong, 2014.

[6] G. Antonelli, F. Caccavale, F. Grossi and A. Marino, "Simultaneous Calibration of Odometry and Camera for a Differential Drive Mobile

Robot," in *IEEE International Conference on Robotics and Automation*, Anchorage, Alaska, 2010.

[7] S. B. Lazarus, P. Silson, A. Tsourdos, R. Zbikowski and A. B. White, "Multiple sensor fusion for 3D navigation for unmanned autonomous vehicles," in *18th Mediterranean Conference on Control & Automation*, Marrakech, Morocco, 2010.

[8] G. Sepulveda, J. C. Niebles and A. Soto, "A Deep Learning Based Behavioral Approach to Indoor Autonomous Navigation," in *IEEE International Conference on Robotics and Automation (ICRA)*, Brisbane, Australia, 2018.

[9] R. Algabri and M.-T. Choi, "Deep-Learning-Based Indoor Human Following of Mobile Robot Using Color Feature," *Sensors,* pp. 1–19, 5 May 2020.

[10] R. Siegwart and I. R. Nourbakhsh, Introduction to Autonomous Robots, Cambridge, Massachusetts: The MIT Press, 2004.

[11] M. Filipenko and I. Afanasyev, "Comparison of Various SLAM Systems for Mobile Robot in an Indoor Environment," in *International Conference on Intelligent Systems (IS)*, Russia, 2018.

[12] K. Krinkin, A. Filatov, A. Filatov, A. Huletski and D. Kartashov, "Evaluation of Modern Laser Based Indoor SLAM Algorithms," in *PROCEEDING OF THE 22ND CONFERENCE OF FRUCT ASSOCIATION*, Petrozavodsk, Russia, April 2018.

[13] D. Fox, W. Burgard, F. Dellaert and S. Thrun, "Monte Carlo Localization: Efficient Position Estimation for Mobile Robots," in *Sixteenth National Conference on Artificial Intelligence*, Orlando, Florida, 1999.

[14] M. Labbe and F. Michaud, "Long-Term Online Multi-Session Graph-Based SPLAM with Memory Management," *Autonomous Robots,* vol. 42, no. August 2018, pp. 1133–1150, 2017.

[15] C. S. Yun, S. Parasuraman and V. Ganapathy, "Dynamic Path Planning Algorithm in Mobile Robot Navigation," in *IEEE Symposium on Industrial Electromics and Applications*, Langkawi, Malaysia, September.

[16] S. Garrido, L. Moreno, M. Abderrahim and F. Martin, "Path Planning for Mobile Robot Navigation using Voronoi Diagram and Fast Marching," in *International Conference on Intelligent Robots and Systems*, Beijing,China, 2006.

[17] O. Brock and O. Khatib, "High-speed Navigation Using the Global Dynamic Window Approach," in *International Conference on Robotics & Automation*, Detroit. Michigan, May, 1999.

[18] A. Kelly and B. Nagy, "Reactive Nonholonomic Trajectory Generation via Parametric Optimal Control," *The International Journal of Robotics Research,* vol. 22, no. July-August 2003, pp. 583–601, 2003.

[19] G.-Z. Yang, B. J. Nelson, R. . R. Murphy, H. Choset, H. Christensen, S. H. Collins, P. Dario, K. Goldberg, K. Ikuta, N. Jacobstein, D. Kragic, R. H. Taylor and M. McNutt , *Combating COVID-19—The role of robotics in managing public health and infectious diseases,* ROBOTS AND SOCIETY, 2020.

[20] H. Z. Khan, A. Siddique and W. C. Lee, "Robotics Utilization for Healthcare Digitization in Global COVID-19 Management," *International Journal of Environmental Research and Public Health,* no. 17, 2020.

[21] R. C. Kovach, Y. Taneli, T. Neiman, M. E. Dyer, J. A. A. Arzaga and . T. S. Kelber, "Evaluation of an ultraviolet room disinfection protocol to decrease nursing home microbial burden, infection and hospitalization rates," *BMC Infectious Diseases,* pp. 1–8, 2017.

[22] A. Hand, "Healthcare Packaging," 2 April 2020. [Online]. Available: https://www.healthcarepackaging.com/covid-19/article/21126536/covid19-provides-use-cases-for-mobile-robotics.

[23] E. Ackerman, "IEEE Spectrum," 27 August 2019. [Online]. Available: https://spectrum.ieee.org/automaton/robotics/industrial-robots/blue-ocean-robotics-acquires-suitable-technologies-beam-telepresence-robot.

[24] K. Wiggers, "The Machine," 19 December 2018. [Online]. Available: https://venturebeat.com/2018/12/19/roboteam-raises-21-million-for-temi-a-home-robot-that-can-follow-you-around/.

[25] S. Crowe, "Robotics Business Review," 3 January 2018. [Online]. Available: https://www.roboticsbusinessreview.com/consumer/roombas_will_help_clean_up_your_homes_weak_wifi/.

[26] "Peanut Robotics," [Online]. Available: https://peanutrobotics.com/.

[27] P. Verma, "The Economic Times," 16 April 2020. [Online]. Available: https://economictimes.indiatimes.com/industry/healthcare/biotech/aiims-taps-robots-and-telemedicine-to-cut-contact-risk/articleshow/75171727.cms?from=mdr.

[28] A. A. MALIK, *Robots and COVID-19: Challenges in integrating robots for collaborative automation,* Researchgate, 2020.

[29] M. E. Romero, "The World," 8 April 2020. [Online]. Available: https://www.pri.org/stories/2020-04-08/tommy-robot-nurse-helps-italian-doctors-care-covid-19-patients.

[30] "The Guardian," [Online]. Available: https://www.theguardian.com/world/2018/feb/06/japan-robots-will-care-for-80-of-elderly-by-2020.

[31] E. Ackerman, "IEEE Spectrum," 27 April 2020. [Online]. Available: https://spectrum.ieee.org/automaton/robotics/medical-robots/akara-robotics-turtlebot-autonomous-uv-disinfecting-robot.

[32] H. Z. Khan, A. Khalid and J. Iqbal, "Towards realizing robotic potential in future intelligent food manufacturing systems," *Innovative Food Science and Emerging Technologies,* pp. 11–24, 2018.

[33] M. Tavakoli, J. Carriere and A. Torabi, *Robotics For COVID-19: How Can Robots Help Health Care in the Fight Against Coronavirus,* Robotic Systems Research @ University of Alberta - Robotics For COVID-19, March 2020.

6

Predictor System for Tracing COVID-19 Spread

Kuldeep Panwar[1], Supriya Pandey[1], Kamal Rawat[2], and Neeraj Bisht[3]

[1]General Electric Power, Noida 201303, India
[2]Department of Mechanical Engineering, Meerut Institute of Engineering & Technology, Meerut 250005, India
[3]Department of Mechanical Engineering, G.B. Pant University of Agriculture Engineering, Pantnagar 263153, India
Corresponding Author: Kuldeep Panwar, kuldeeppanwar.kec@gmail.com.

Abstract

According to World Health Organization (WHO), the first case of COVID-19 inflicted pneumonia was reported in Wuhan in December 2019; today, not only china which was the center of this epidemic in the early stage but the whole world is affected with this virus. In the absence of any vaccine or treatment, the ferocity of any epidemic can be measured only by predicting its transmission among the population. Various evidences of human-to-human transmission by many clinical studies on COVID-19 have shown that its transmission rates are much higher than similar coronavirus such as severe acute respiratory syndrome (SARS) coronavirus and Middle East respiratory syndrome (MERS) coronavirus. To predict the rate of transmission from an infected person into the local population in case of a pandemic like COVID-19, effective reproduction number (R_0) is an important parameter. Effective reproduction number is an indication of transmissibility of the virus in a native population. It represents average number of new infections caused by a previously infected person. If the

79

value of R_0 is higher than 1, it means transmission of virus increases, and with R_0 less than 1, transmission rate reduces. There are various mathematical models to predict effective reproduction number (R_0) for a pandemic. The current chapter discusses various mathematical models based on susceptible exposed infectious recovered death cumulative (SEIRDC) structure.

6.1 Introduction

According to the timeline of World Health Organization (WHO), in December 2019, Wuhan Municipal Health Commission reported first few cases of COVID-19 in Wuhan in Hubei Province of China [1]. The first case reported outside China was on 13 January 2020 in Thailand. As of 8 June 2020, COVID-19 virus has spread in almost all the countries with total 6,931,000 cases and 400,857 deaths [2]. In the absence of vaccine, the only effective measure to tackle the epidemic has been lockdowns. Almost all countries in Europe and South East Asia imposed lockdowns in their countries. India soon followed the suite and a nationwide lockdown was imposed from 24 March 2020. After four phases of lockdown, unlock of various activities has begun, and the total number of cases in the period has gone up to 246,109 with 7200 number of deaths [2]. Various clinical studies on COVID-19 have shown evidences of human-to-human transmission similar to severe acute respiratory syndrome (SARS) coronavirus and Middle East respiratory syndrome (MERS) coronavirus. In these cases, effective reproduction number (R_0) is an important parameter. Effective reproduction number is an indication of transmissibility of the virus in a native population. It represents average number of new infections caused by previously infected person. If the value of R_0 is higher than 1, it means transmission of virus increases, and with R_0 less than 1, transmission rate reduces.

6.2 Various Prediction Methods

There are lots of studies done to predict the rate of spread of COVID-19 virus in native population. In this chapter, we have identified 12 such studies which can estimate the basic reproductive number for COVID-19. Table 6.1 shows various studies identified to predict the effective reproduction number for COVID-19.

In the next section, a case study is discussed to predict the infection rate with the help of one of the above-discussed methods. Several scientists have developed various mathematical models to predict effective reproduction

Table 6.1 Calculated values of K for different lockdown phases.

study	Method	Approach
Joseph et al.	Stochastic Markov Chain Monte Carlo methods (MCMC)	MCMC methods with Gibbs sampling and non-informative flat prior, using posterior distribution.
Shen et al.	Mathematical model, dynamic compartmental model with population divided into five compartments: susceptible individuals during the incubation period, infectious individuals with symptoms, isolated individuals with treatment and recovered individuals	$R_0 = \beta/\alpha$; β = mean person-to-person transmission rate/day in the absence of control interventions, using nonlinear least squares method to get its point estimate; α = isolation rate = 6
Liu et al.	Statistical exponential growth using SARS generation time = 8.4 days, SD = 3.8 days	Applies Poisson regression to fit the exponential growth rate $R_0 = 1/M(-r)$; M = moment generating function of the generation time distribution; r = fitted exponential growth rate
Liu et al.	Statistical maximum likelihood estimation, using SARS generation time = 8.4 days, SD = 3.8 days	Maximize log-likelihood to estimate R_0 by using surveillance data during a disease epidemic, and assuming the secondary case is Poisson distribution with expected value R_0
Read et al.	China 1–22 January 2020 Mathematical transmission model assuming latent period = 4 days and near to the incubation period	Assumes daily time increments with Poisson-distribution and apply a deterministic SEIR metapopulation transmission model, transmission rate = 1.94, infection period = 1.61 days
Majumder et al.	Mathematical Incidence Decay and Exponential Adjustment (IDEA) model	Adopted mean serial interval lengths from SARS and MERS ranging from 6 to 10 days to fit the IDEA model
Cao et al.	Mathematical model including compartments	Susceptible-Exposed-Infection- Recovered-Death-Cumulative (SEIRDC) $R = K2 (L \times D) + K(L + D) +1$ L = average latent period = 7, D = average latent infectious period=9, K=logarithmic growth rate of the case counts
Zhao et al.	Statistical exponential growth model method adopting serial interval from SARS (mean = 8.4 days, SD = 3.8 days) and MERS (means = 7.6 days)	SD = 3.4 days) corresponding to 8-fold increase in the reporting rate $R_0 = 1/M(-r)r$ = intrinsic growth rate M = moment generating function
Imani	Mathematical model, computational modelling of potential epidemic trajectories	Assume SARS-like levels of case-to-case variability in the numbers of secondary cases and a SARS-like generation time with 8.4 days, and set number of cases caused by zoonotic exposure and assumed total number of cases to estimate R0 values for best-case, median and worst-case
Julien and Althaus	Stochastic simulations of early outbreak trajectories	Stochastic simulation of early outbreak trajectories were performed that are consistent with the epidemiological finding to date
Tang et al.	Mathematical SEIR –type epidemiological model incorporates appropriate compartments corresponding to interventions	Method-based method and Likelihood-based method
Qun Li et al.	Statistical exponential growth model	Means incubation period = 5.2 days, mean serial interval = 7.5 days

number (R_0) for a pandemic. Read *et al.* developed a model assuming daily time increment with Poisson-distribution taking latent period of four days and infectious period of 1.61 days [6]. Imai *et al.* in their model assumed levels similar to SARS, with a generation time of 8.4 days and used total number of cases to calculate best and worst values of R_0 [7]. Tang *et al.* developed susceptible exposed infectious recovered (SEIR) type model [8]. Cao *et al.* developed a mathematical model based on susceptible exposed infectious recovered death cumulative (SEIRDC) structure. The average latent period (L) and average latent infectious period (D) were taken as 7 and 9, respectively [5]. The sensitivity of model was also checked by adopting different values of L and D.

6.3 Case Study – Prediction of Effective Reproductive Number for India

In the current case study, to understand the transmissibility of COVID-19 virus, SEIRDC model is used to predict accurate values of effective reproduction number (R_0) for the period of 25 March 2020 to 1 June 2020. In the current chapter, a comparative study of R_0 is done for all the phases of lockdown imposed by India to check the effects of lockdown on the transmissibility of the virus in the country.

To understand the transmissibility of COVID-19 virus and predict accurate values of effective reproduction number (R_0), the SEIRDC model is used in the current paper. For the calculation of R_0, average infection period (D) is taken as 3.96 and the average latent period (L) is taken as 3 [9]. The equation used is as follows:

$$R_0 = K^2(L \times D) + K(L + D) + 1 \tag{6.1}$$

where
R_0 is effective reproduction number;
K is logarithmic growth rate of infected cases in India;
L is the average latent period;
D is the average latent infectious period.

K is calculated on the basis of number of days since the first positive case in India occurred and for different time points according to the phases of lockdown. Table 6.2 shows the different phases of lockdown with number of days in each phase and calculated value of K for each phase.

Table 6.2 Calculated values of K for different lockdown phases.

S. No	Lockdown Phase	Start Date	End Date	Total No. of Days	K
1	Phase 1	25 March 2020	14 April 2020	21	0.225
2	Phase 2	15 April 2020	3 May 2020	19	0.151
3	Phase 3	4 May 2020	17 May 2020	14	0.128
4	Phase 4	18 May 2020	31 May 2020	14	0.112
5	Phase 5	1 June 2020	30 June 2020	30	0.105 (till 8 June)

6.4 Results and Discussions

India imposed lockdown in phases to prevent the spread of COVID-19 virus. Phase 1 (21 days) included nation-wise suspension of all services, transport, and factories. In Phase 2 (19 days), the country was divided into three zones, namely Red zone (high infection cases area), Orange zone (low infection cases areas), and Green zones (zero cases). There were some relaxations provided in Orange and Green zone areas, while Red zones were completely locked down. In Phase 3 (14 days), further relaxations were given in Orange and Green zones, while Red zones remain under lockdown. In Phase 4 (14 days), with additional relaxation, Red zones were further categorized into containment zones and buffer zones.

In Phase 5 (30 days), lockdown restrictions are imposed only in containment zones. Figure 6.1 shows the number of daily cases during all phases of lockdown and Figure 6.2 shows the cumulative number of cases.

Phase 1 – Phase 1 started on 25 March 2020 with total cases 552 (29 March) and ended with total 9844 cases (14 April). Figure 6.3 shows calculated effective reproduction number (R_0) on 29 March as 4.6 which means an infected person infects more than four people, which is also higher than 2003 SARS epidemic ($R_0 = 4.0$) [10]. R_0, by the end of phase 1, decreases to 3.17, which means an infected person now infects three people during this outbreak. This is a clear indication that complete lockdown prevented the epidemic from spreading at the rate at which it transmitted in Wuhan [5].

Phase 2 – During Phase 2, the effective reproduction number (R_0) further decreased to 2.34 in which case one infected person can infect two people. During both Phases 1 and 2, the transmissibility of virus was controlled, which can be seen as the decrease in R_0.in Figure 6.3.

Phase 3 – During this phase, the number of cases increased from 42,014 (4 May) to 90,408 (17 May), with a decrease in R_0 from 2.34 to 2.09, which means the numbers of people infected from one person remains the same as

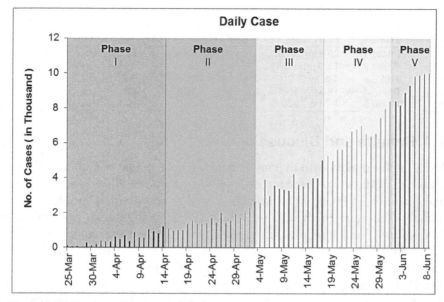

Figure 6.1   Number of cases reported daily during different phases of lockdown in India.

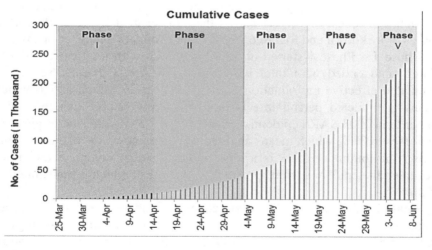

Figure 6.2   Cumulative number of cases reported during different phases of lockdown in India.

in Phase 2. This could be due to various relaxations provided in this phase of lockdown.

  Phase 4 – At the end of this phase of lockdown, the value of R_0 is 1.93 (31 May). Figure 6.3 shows that there is very small change in this valve on 18 May, which is 2.09, and on 31 May, it is 1.93.

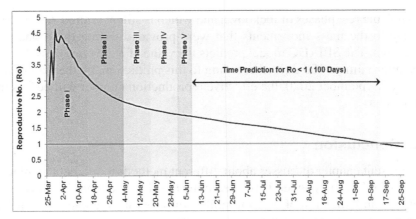

Figure 6.3 Effective reproduction number (R_0) during lockdown phases in India.

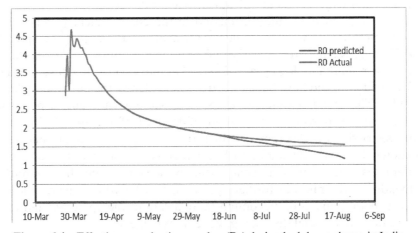

Figure 6.4 Effective reproduction number (R_0) during lockdown phases in India.

From Figure 6.2, as of 8 June 2020, the R_0 for India is calculated as 1.86, with total number of cases being 246,109 out of which 124,429 cases have been cured.

Figure 6.4 shows a comparison between the actual and the predicted values of effective reproductive number. It is evident from the phase-wise comparison of lockdown in India that R_0 decreased from an alarming value of 4.6 (29 March 2020) to 1.86 (8 June 2020). To have a total control over this pandemic, R_0 is to be reduced to less than 1.

It was also seen that during the first two phases of lockdown, India was able to control effective reproduction number for COVID-19 from 4.6 to 2.3 which is almost 50% decrease in R_0. Whereas, this decrease of R_0 was slowed down in the rest phases of lockdown and which is still at around 2. This could be due to the mass movements that were provided during these phases of lockdown. The SEIRDC model predicts the value of R_0 to come less than 1 in next approximate 100 days. According to this prediction, by the start of third week of September 2020, the effective reproduction number will be less than 1 for India.

6.5 Conclusion

The present chapter discusses about different methods to predict the infection rate for COVID-19 pandemic by predicting the effective reproduction number which is an indication of transmissibility of the virus in a native population. R_0 represents average number of new infections caused by previously infected person. If the value of R_0 is higher than 1, it means transmission of virus increases and with R_0 less than 1 transmission rate reduces.

The results of case study discussed in the chapter clearly indicate that the complete lockdown was a successful decision to contain the spread of COVID-19. As compared to early rate of transmissibility in Wuhan, the first phase saw remarkably lower rate of spread in India. The first phase saw a rate of 3.17 which would have been disastrous considering the population density of India. The lockdown created a discipline among the general population which can be seen from the further rate of decrease of transmissibility despite some restrictions being removed in further lockdowns. At the end of the lockdown 4, the rate decreased to 1.93 which is satisfactory compared to the earlier rates of transmission. It can further be seen that even after the removal of lockdown, the rate has further decreased to 1.83 which clearly indicates to the general population becoming more aware of the epidemic and taking right actions like social distancing, general hygiene, etc.

References

[1] Coronavirus Disease (COVID-19) –WHO Timeline Covid-19. Available at: https://www.who.int/news-room/detail/27-04-2020-who-timeline---covid-19[accessed June 8, 2020].

[2] Coronavirus Update (live): Available at: https://www.worldometers.info/coronavirus [accessed June 8, 2020].

[3] T. Liu, J. Hu, M. Kang *et al.* Transmission dynamics of 2019 novel coronavirus (2019-nCoV). *bioRxiv* 2020. doi: https://doi.org/10.1101/2020.01.25.919787.

[4] Z Zhang Cao, X. Lu *et al.* Estimating the effective reproduction number of the 2019-nCoV in China. *medRxiv* 2020. doi:https://doi.org/10.1101/2020.01.27.20018952.

[5] J. Riou,CL, Althaus. Pattern of early human-to-human transmission of Wuhan 2019-nCoV. *bioRxiv* 2020 https://www.biorxiv.org/content/10.1101/2020.01.23.917351v1.full.pdf.

[6] JM Read, JRE Bridgen, DAT Cummings , A Ho , CP Jewell. Novel coronavirus 2019-nCoV: early estimation of epidemiological parameters and epidemic predictions. medRxiv 2020. doi: https://doi. org/10.1101/2020.01.23.20018549.

[7] N Imai , A Cori, I Dorigatti et al. Transmissibility of 2019- nCoV. 2020. WHO Collaborating Centre for Infectious Disease Modelling, MRC Centre for Global Infectious Disease Analysis, J-IDEA, Imperial College London, UK.

[8] B Tang , X Wang , Q Li et al. Estimation of the transmission risk of 2019-nCov and its implication for public health interventions (January 24, 2020). https://ssrn.com/abstract=3525558 or https://doi.org/10.2139/ssrn.3525558 (9 February 2020, date last accessed).

[9] M. Yinon, et al. Science Forum: SARS-CoV-2 (COVID-19) by the numbers. Elife, 2020, 9: e57309. DOI: 10.7554/eLife.57309

[10] Y. Liu Y, A. A. Gayle, Wilder-Simth A, Rocklov J, The reproductive number of COVID-19 is higher compared to SARS coronavirus. Journal of Travel Medicine, 2020, 1–4 doi: 10.1093/jtm/taaa021.

[11] L. Qun et al., Early transmission dynamics in wuhan, china of novel coronavirus-infected pneumonia. New England Journal of Medicine, 2020, 1056/NEJMoa2001316.

[12] S. Zhao, J. Ran, ss. Musa et al., Preliminary estimation of the basic reproduction number of novel coronavirus in china from 2019 to 2020: a datadriven analysis in the early phase of the outbreak, 2020, bioRxiv 2020, doi: 10.1101/2020.01.23.916395.

[13] Z. Cao Zhang, X Lu et al. Estimating the effective reproduction number of the 2019-ncoV in China. medRxiv 2020, https://doi.org/10.1101/2020.01.27.20018952.

[14] M. Majumdar, KD. Mandal, Early transmissibility assessment of novel coronavirus in Wuhan, China, 2020, https://paper.ssrn.com/abstract=3524675.

[15] M. Shen, Z. Peng, Y. Xiao, L. Zhang, Modelling the epidemic trend of the 2019 novel coronavirus outbreak in china. bioRxiv 2020, https://doi.org/10.1101/2020.01.23.916726.

7

Discovery of Robust Distributions of COVID-19 Spread

Chhaya Kulkarni, Sandipan Dey, and Vandana Janeja

Department of Information Systems, University of Maryland, Baltimore County, Baltimore, MD, USA
Corresponding Author: Chhaya Kulkarni, Vandana Janeja, ckulkar1@umbc.edu, sandipan.dey@gmail.com, vjaneja@umbc.edu.

Abstract

COVID-19 as a pandemic has impacted many lives with continued threat to our way of life. Decision-makers are grappling with the spread without insights into whether they are in a peak or a plateau of the spread. This severely limits resource management in regions severely affected by COVID-19. Analysis of COVID-19 cases and deaths as time-series distributions can yield insights that can aid the efforts in such regions to help estimate the curve of the spread. In this chapter, we use time-series analysis to analyze COVID-19 spread at various locations to study the overall patterns that emerge. We utilize methods including piecewise aggregate approximation (PAA), matrix profiles (MPs), and time-series discretization to identify time periods where the number of cases and deaths reported depicted any major anomalies and where the overall time series seems to be following a trend. Our analysis can prove beneficial to understand the distributions of COVID-19 cases and deaths to quantify the data into different trends to show when the number of cases spiked and when they remained fairly consistent, even when they are in a trend of high number of cases and deaths. This type of trend analysis is particularly useful to compare locations with similar metadata such as number of assisted living communities, population densities, etc., and study if the disease spread is similar or deviating. This can also provide insights into how the policies have had an impact on the spread at these locations.

7.1 Introduction

COVID-19 has impacted the world at large and caused immense loss of life. It is vital for decision-makers to take steps now more than ever to help contain further spread. In this chapter, we propose a method to identify peaks and trends in COVID-19 data that can help discover when the number of cases and deaths has spiked substantially and when the cases have risen and/or remained steady for a while. Analyzing these patterns can generate insights about COVID-19 to help decision-makers take informed decisions. This information can help further probe into the cause of this kind of pattern and mitigate the further spread of COVID-19. We propose a way by which we can generate distributions to represent patterns of spread across locations, which can help create a potential searchable database of distributions by geospatial features. This can help decision-makers identify similar trends for their own jurisdictions from a historical database of distributions.

In this chapter, we have used a combination of approaches to evaluate COVID-19 data as a time series. Temporal data analysis can help in comparing different locations and assessing what the similarities and differences between those locations are and how certain policies may impact the outcomes of the spread.

We utilized the data from the New York Times [11] and applied temporal data analysis including traditional outlier detection, matrix profiles (MPs), piecewise aggregate approximation (PAA), and time-series discretization. Using these methods, we evaluate the outlying spikes in the spread and distributions of the overall spread. These distributions are fairly robust to the day-to-day changes and present the overall picture of the spread. This can be a quick snapshot for analyzing a location's spread as compared to others. Our overall contributions include the following:

i. We provide a method to discover distributions in the COVID-19 data using temporal neighborhood discovery.
ii. We utilize outlier detection methods on the COVID-19 cases and death data to identify unusual days in terms of the spikes or changes in the disease spread.
iii. We also provide an analysis of COVID-19 distributions across various geospatial locations with their metadata features such as population density or types of populations in those regions.

The rest of the chapter is organized as follows. In Section 7.2, we discuss our approach in detail. In Section 7.3, we illustrate the experimental results across multiple locations. Finally, we discuss our overall findings and future work in Section 7.4.

7.2 Methodology

In this work, we explore how COVID-19 has spread across various locations over time. Here the geospatial locations may have similar features such as population density, number of assisted living facilities, or other demographic features. However, despite the feature similarity of the geospatial locations, the spread of the disease may be different based on the policies that were followed in terms of social distancing or availability of resources in the health care operations. We aim to derive representations of the COVID spread in terms of a distribution over time. In addition, we also explore time periods when the spread is very unusual as compared to the rest of the distribution at a particular location. In this chapter, we aim to show distributions within the temporal data of different regions where COVID-19 has spread with specific trends over time. Essentially, we have accomplished two major tasks, namely outlier detection and identification of distributions with similar and dissimilar characteristics in the geospatial regions. COVID-19 raw data from NYTimes [11] was available cumulatively across days. We preprocessed the data to transform into non-cumulative data. Outlier detection and neighborhood detection techniques were applied on the processed data. Outlier detection was performed using two techniques including standard, statistical methods and isolation forest [4]. We utilized methods such as PAA [3] and MPs [2] to discover similar behaviors across the time period. We perform outlier detection on the outcomes from these two methods as well to contrast similarities and deviations across locations. The overall distribution of the spread of COVID-19 was discovered using temporal neighborhood discovery and time-series discretization methods using the SMerg algorithm [1]. Outlier detection and identifying distributions across locations and times may help assess the spread of COVID-19 across various geospatial locations to find similar or dissimilar trends. Here and henceforth by (temporal) distribution we shall simply refer to the time series data. If we consider time as a random variable T, then each value (i.e., the number of cases) corresponding to a particular time instant $T=t$ can be represented as $n(T=t)$, which can be thought of as a realization of a temporal histogram or (un-normalized) temporal distribution. The overall approach is shown in Figure 7.1.

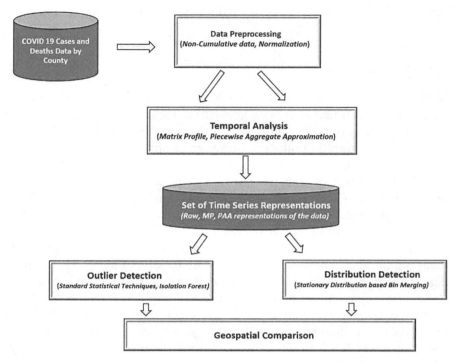

Figure 7.1 Overall methodology.

7.2.1 Data Preprocessing

We performed basic data preprocessing to get the data ready for the various analysis and algorithmic inputs. We also performed consistent normalization across all data streams to take away major scale fluctuations to align all cases and deaths between 0 and 1. For data selection, we focused on key geospatial locations with their metadata or features that helped us understand the similarities and dissimilarities. Most data about cases and deaths was aggregated over days; however, we wanted to see the actual cases and deaths for each day, and for this, we derived the non-aggregated data for individual time points.

7.2.2 Temporal Analysis

Temporal analysis helps to observe the behavior of a variable in a particular scenario over time. We used methods such as MPs and PAA for an understanding of the variations including similarities and deviations in the temporal data. MP shows the time periods where there are largest changes in data values. PAA, on the other hand, helps in visualizing the data in a reduced dimensionality. In a way, PAA also provides the mean distribution in the time-series data. We have employed these techniques on COVID-19 data to understand the way in which COVID-19 has progressed across different locations in the United States. We next describe these methods in brief.

Matrix Profiles: MP uses motifs to describe similar patterns and discords for outliers. MP divides the temporal data into subsequences (windows). Each subsequence is compared with other subsequences in the time-series data. Euclidean distance between the subsequences is recorded in a structure called distance profile. Distance profile always records the minimal distance between two subsequences. Furthermore, a profile index structure is created, which contains the index of the most similar subsequences. Primary advantages of using MP are that it is easily scalable and domain-agnostic.

PAA: PAA is a well-known dimensionality reduction technique in time-series mining. It is popular for being able to visualize the data in a condensed form. Input to the PAA algorithm is a time series. PAA divides the time series into a set of segments, and each segment is replaced by the mean of its data points. PAA proves to be helpful in detecting the trend in spread of COVID-19 cases and deaths. For example, Figure 7.2 shows PAA representation on raw data for COVID-19 cases in Baltimore City.

Figure 7.2 illustrates the analysis over COVID-19 cases reported in Baltimore City in the state of Maryland: (a) PAA applied over raw data for COVID-19 cases and (b) MP of the data. The PAA depicts the mean distribution over the time series and the MP spikes identify the time periods where there are major changes occurring, including the drops or spikes around that time period.

7.2.3 Distribution Detection

In order to find distributions in temporal data, we have used our prior work on discovering temporal neighborhoods [1]. We have used this technique to discretize the COVID-19 data into temporal neighborhoods based on similarity in terms of number of cases or deaths. This will help us in comprehending which locations demonstrate a similar or different pattern depicting the effect by COVID-19. This technique is based on two algorithms, namely Similarity based Merging (SMerg) and Markov stationary distribution based Merging (StMerg). Both the algorithms follow the Markov model. Figure 7.3 illustrates the process adopted in SMerg and StMerg algorithms.

We start by initially dividing the COVID-19 temporal data into equal depth bins. These equal-depth bins are treated as the states of the Markov model. We then compute the distributional distance in between each bin and

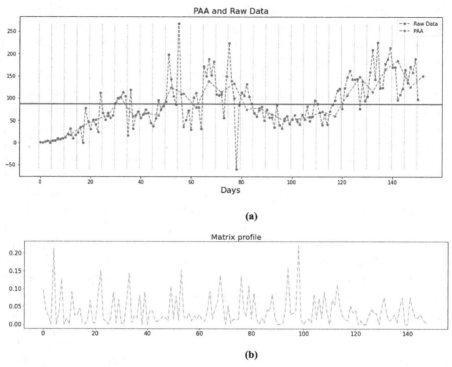

Figure 7.2   Summarization of the time-series data for Baltimore City. (a) PAA on raw data for COVID-19 cases in Baltimore City. (b) Matrix profile for COVID-19 cases in Baltimore City.

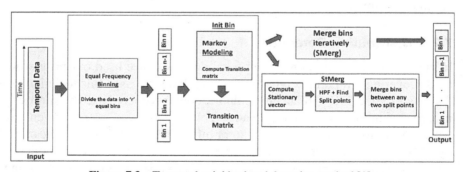

Figure 7.3   Temporal neighborhood detection method [1].

its neighboring bins using different distance metrics such as Mahalanobis, Kullback-Leibler distance measure (KL), and Bhattacharyya [8]. Next we create the Markov transition matrix based by computing a row-stochastic similarity matrix from the distance matrix. From the transition matrix, the adjoining bins with high degree of similarity can be merged to generate the temporal neighborhood. In our experiment, we have mainly utilized the SMerg algorithm. The STMerg algorithm computes the Markov steady-state vector (the stationary distribution) and uses a high-pass filter on the steady state vector to detect changes, and accordingly partitions the temporal data with the final unequal depth discretized bins.

We carried out the temporal neighborhood generation for each of the five counties including Baltimore County, Baltimore City, Montgomery County, Prince George's County, and Howard County in the state of Maryland for analyzing both cases and deaths by COVID-19, respectively. We selected these locations because we are interested in seeing how COVID-19 has impacted populations in similar kinds of geospatial locations. Another reason was to analyze whether COVID-19 has left a trail of similar or dissimilar patterns in these counties. Discovering these patterns would help identify the impact of policies implemented at these locations, to gain insights into preventing further spread of COVID-19.

7.2.4 Outlier Detection

7.2.4.1 Temporal Outlier Detection

We represented the COVID-19 cases and spread using the raw daily numbers. We used a time-series method, namely the MPs [2]. MP helps discover similarities and outliers in temporal data. We wanted to be able to identify not only the spikes in the raw data but also time periods where the trends were very similar or dissimilar across the time period for a particular location. Hence, we used MP structure in our experiment.

In addition to the MP and raw data representation, we also used PAA [3] on both raw data and MP representation of the data.

We computed the mean and standard deviation (SD) for both cases and deaths in each of the locations from the MP array of values for COVID-19 cases and deaths. We then computed outliers based on the deviations in terms of spread around the mean behavior and in terms of the SDs from the mean. As expected, the one SD produces the deviations to a large degree as compared to the 3SD.We captured the accuracy and precision values as compared to the ground truth peaks we observed in the data in terms of unusual spreads of the disease. In addition to computing the outliers based on the standard statistical techniques, we also used **isolation forest** [4]. Isolation forest is fast in terms of execution time and also requires less memory. Another advantage of isolation forest is that it is an unsupervised learning algorithm. Our experiment is also in the unsupervised domain; therefore, we decided to apply the isolation forest algorithm on our processed COVID-19 data.

Traditional anomaly detection algorithms work by defining normal behavior and any data point that deviates from this normal behavior is identified as an anomaly. Isolation forest is an ensemble method that incorporates a decision tree algorithm. It recursively generates partitions on the dataset by randomly selecting a feature and then randomly selecting a split value for the feature. The anomalies need fewer random partitions to be isolated compared to "normal" points in the dataset; so the anomalies will be the points which have a smaller path length in the tree, path length being the number of edges traversed from the root node in the decision tree [14].

7.3 Experimental Results

7.3.1 Geospatial Context of the Data

NYTimes COVID-19 dataset was used in our experiment [11]. This data was updated on a daily basis. This data was also compared with Maryland Government's COVID-19 data. Data that we have used in our experiment was largely related to the number of COVID-19 cases and deaths reported in five different counties in Maryland State. We selected Prince George's, Baltimore, Baltimore City, Howard, and Montgomery County. We also wanted to contrast with another location with a very high density population and large number of assisted living facilities; so we included New York City as it had extreme cases at one point of time. We selected these locations to get a variable set of counties with different types of rates of spread. Prince George's, Montgomery, Baltimore county, and Baltimore city were some of the counties in Maryland that were impacted much more than other counties in terms of the number of cases and deaths. We also looked at some of these counties with diverse populations. Our data is from the dates of 29 February 2020 to 12 August 2020. At present, in Maryland, Prince George's county has the leading number of COVID-19 cases and deaths. Both Baltimore county and Baltimore city have a greater number of senior citizens [13]. We also evaluated the features, as shown in Table 7.1, of these locations to perform comparative analysis. Our data was derived from demographics data collected by United States Census Bureau [17] and [18].

7.3.2 Results

Our results include a discussion of the methods used, namely standard statistical techniques, PAA, MP, and isolation forest for outlier detection. The implementations for PAA [25], MP [24], and isolation forest [14] were used.

Table 7.1 Demographics Data.

County name	Population	Population of senior citizens	Number of assisted living facilities	Number of hospitals	% of population distribution	Population density per square mile	Household income/ income distribution
Prince George's	909,327	177,327	8	7	White – 12.3 Black – 64.4 Latin – 19.5 Asian – 4.4 Mixed – 2.7 American Indian – 1.2 Hawaiian – 0.2	1788.8	$81,969
Baltimore City	593,490	116,857	14	17	White – 27.7 Black – 62.7 Latin – 5.7 Asian – 2.7 Mixed – 2.2 American Indian – 0.5 Hawaiian – 0.1	7671.5	$48,840
Baltimore County	827,370	206,565	20	5	White – 55.8 Black – 30.3 Latin – 5.8 Asian – 6.3 Mixed – 2.6 American Indian – 0.4 Hawaiian–0.1	1345.5	$74,127

Continued

Table 7.1   Continued

County name	Population	Population of senior citizens	Number of assisted living facilities	Number of hospitals	% of population distribution	Population density per square mile	Household income/ income distribution
Montgomery	1,050,688	235,193	8	11	White – 42.9 Black – 20.1 Latin – 20.1 Asian – 15.6 Mixed – 3.5 American Indian – 0.7 Hawaiian – 0.1	1978.2	$106,287
Howard County	325,690	71,364	7	1	White – 50.3 Black – 20.4 Latin – 7.3 Asian – 19.3 Mixed – 3.9 American Indian – 0.4 Hawaiian – 0.1	1144.9	$117,730
New York City	8,336,817	1,175,491	78	67	White – 32.1 Black – 24.3 Latin – 29.1 Asian – 13.9 Mixed – 3.5 American Indian – 0.4 Hawaiian – 0.1	27,012.5	$60,762

We have also shown the results for distributions discovered using the SMerg algorithm.

7.3.2.1 Temporal Analysis and outlier Detection

We computed the mean and SD for both cases and deaths in each of the five counties from the MP values for COVID-19 cases and deaths. We then calculated values deviating around the central tendency. We did this to compute the outliers across 1, 1.5, 2, and 3 SDs around the mean.

We also evaluate the outliers we found as compared to ground truth based on the spread of the disease in the various counties. We discuss results for Baltimore County. Our ground truth was based on findings from additional external sources [19–21].

Figure 7.4 shows the values for accuracy across the methods. It can be seen that PAA has the highest precision followed by MP. We also notice that MP is better suited for detecting changes in the distribution rather than global outliers in this case.

We found interesting results using the isolation forest algorithm. The representation in Figure 7.5 shows the days when either the highest or lowest number of COVID-19 cases was recorded in Prince George's County of Maryland. Prince George's County is worst affected by COVID-19. On 1 May 2020 and 24 May 2020, as high as 693 cases were reported.

In Prince George's County, there were 157 days of data. This data was fed into the isolation forest algorithm and the contamination factor was set to 10%; therefore, a total of 15 outliers were detected. We can see that isolation

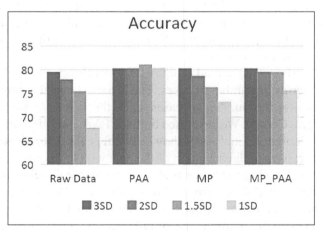

Figure 7.4 Accuracy measures for Baltimore County.

	date	New Daily Cases	scores	anomaly
11	3/20/2020	8	-0.010014	-1
15	3/24/2020	16	-0.048325	-1
16	3/25/2020	13	-0.047590	-1
17	3/26/2020	25	-0.039451	-1
47	4/25/2020	392	-0.068852	-1
53	5/1/2020	693	-0.210521	-1
59	5/7/2020	381	-0.034822	-1
60	5/8/2020	386	-0.015675	-1
66	5/14/2020	377	-0.063745	-1
67	5/15/2020	343	-0.092312	-1
71	5/19/2020	632	-0.241915	-1
73	5/21/2020	385	-0.012294	-1
76	5/24/2020	693	-0.210521	-1
80	5/28/2020	415	-0.155629	-1
102	6/19/2020	34	-0.040649	-1

Figure 7.5   Isolation forest output (Prince George's County).

forest has fetched all those days when the greatest number of cases as well as the day when few cases were reported.

Figure 7.6 shows the PAA and MP for cases in Baltimore County, Maryland. We can observe here that PAA reduces the dimensionality of the data and is helpful in a quick analysis and representation of how COVID-19 has spread. Both PAA and MP show that there was a sudden change in cases in Baltimore County around the 100th day, specifically a drop in cases. On further analysis of data around the 100th day, we can see that there is a major drop from around the 80th day.

Figure 7.7 shows the spread of COVID-19 in Prince George's County in Maryland in terms of number of cases. The *X*-axis shows the number of days and *Y*-axis shows the number of cases since the first case was reported in Prince George's County. We have used both the PAA and MP to track COVID-19 in Prince George's County. MP shows several peaks that represent discords in data; these represent the deviation of the data with respect to other points in the time series. So a high value or peak represents that the distribution at that point is substantially different from the rest of the data. PAA representation aids in understanding the major changes in the MP without losing any knowledge of the fundamental distribution (rise or fall in cases). Prince George's County has been leading in terms of the number of COVID-19 cases reported. Prince

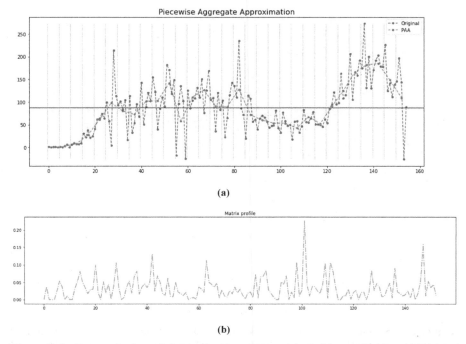

Figure 7.6 Summarization of time-series data for cases in Baltimore County. (a) PAA on raw data for COVID-19 cases in Baltimore County. (b) Matrix profile for COVID-19 cases in Baltimore County.

George's County had the highest number of COVID-19 cases recorded on 1 May 2020 as well as on 24 May 2020 at 693.

Figure 7.8 shows the number of deaths in Prince George's County. It can be seen that the number of deaths were at an all-time high rate between 50th and 55th day [15]. The number of deaths reported continued to be on the rise. Toward the quarter of Figure 7.8(a), the number of deaths has plummeted.

7.3.2.2 COVID-19 Distributions

Figure 7.9 shows distribution of cases in Prince George's County using the temporal discretization algorithm, SMerg. The *X*-axis shows the number of days and the *Y*-axis shows the number of cases. We were able to track the distribution for 157 days. We have tried to show the distribution of cases in the form of plateaus. The overall distribution depicted in a visualization helps to see when the cases are in a steady state even when the overall number of cases is high, for example between the 70th and 80th day [15]. The number of cases dropped significantly after the 80th day.

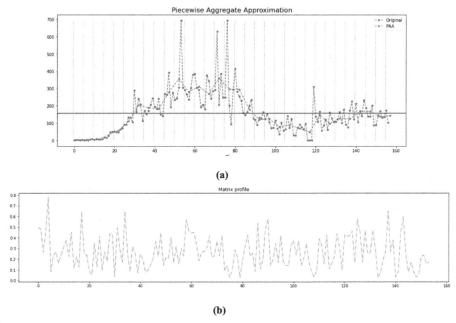

Figure 7.7   Summarization of time-series data for cases in Prince George's County. (a) PAA on raw data for COVID-19 cases in Prince George's County. (b) Matrix profile for COVID-19 cases in Prince George's County.

Figure 7.10 shows the distribution of deaths in Prince George's County. In this plot, we show the plateaus discovered in the distribution of deaths. Analyzing the plateaus shows that deaths in Prince George's County were high from the beginning of the 20th day till the 90th day.

Experiments conducted to trace the spread of COVID-19 across five different geospatial locations in the form of temporal data distributions revealed insights on when there was a rapid rise and fall in COVID-19 cases and deaths and a steady state even when there was a large number of cases and deaths. From the distributions, it is clear that COVID-19 was at its deadliest in the period of late April and all throughout May.

7.3.2.3 Geospatial Comparison

Table 7.1 helps identify which locations are similar in terms of their features such as population, demographic distribution, population density, population of elderly, and household income to name a few. We observed that Prince George's County, Montgomery County, and Baltimore County are roughly similar in terms of population. These three counties have been severely affected by COVID-19 in the state of Maryland. Prince George's

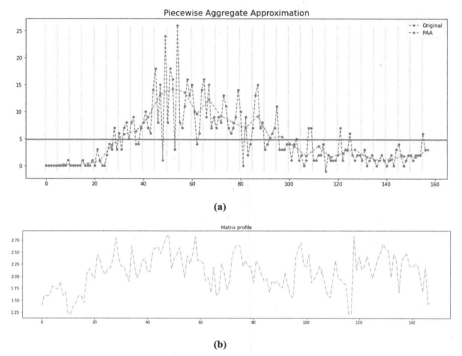

(a)

(b)

Figure 7.8 Summarization of time-series data for deaths in Prince George's County. PAA on raw data for COVID-19 deaths in Prince George's County. Matrix profile for COVID-19 deaths in Prince George's County.

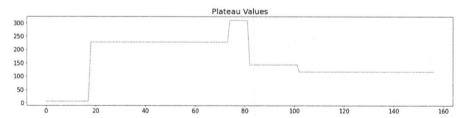

Figure 7.9 Distribution in Prince George's County cases (*initial number of bins: 157; number of partitioned bins: 25; distance measure: Mahalanobis; threshold: 0.20; number of merged bins: 5*).

County has been the highest in terms of the number of cases and deaths reported.

Figure 7.11(a) and (b) both show the COVID-19 distributions for cases and deaths in Montgomery County in Maryland State of the United States, respectively. Montgomery County was the first county in Maryland to report a confirmed case of COVID-19 on 5 March 2020. We collected COVID-19 cases and deaths data for 161 days in Montgomery County. We obtained the

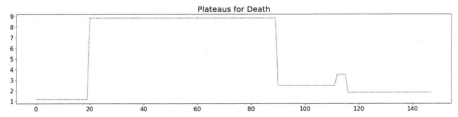

Figure 7.10   Distribution in Prince George's County deaths (*initial number of bins: 157; number of partitioned bins: 25; distance measure: Mahalanobis; threshold: 0.15; number of merged bins: 5*).

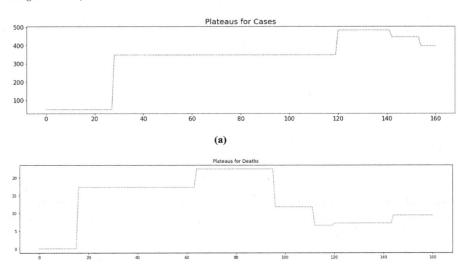

Figure 7.11   COVID-19 distributions for Montgomery County cases and deaths. (a) Distribution of COVID-19 cases (*initial number of bins: 161; number of partitioned bins: 20; distance measure: Mahalanobis; threshold: 0.15; number of merged bins: 5*). (b) Distribution of COVID-19 deaths (*initial number of bins: 161; number of partitioned bins: 20; distance measure: Mahalanobis; threshold: 0.05; number of merged bins: 7*).

distributions for cases and deaths using the SMerg algorithm. The figure also includes the configuration of the parameters used with the SMerg algorithm.

Figures 7.12 and 7.13 both illustrate the distributions for COVID-19 cases and deaths in Baltimore County in the state of Maryland. Both the figures vary in terms of parameters applied to the SMerg algorithm. This experiment with varying parameters was included to understand which distributions were closer to the real-world data. We included 155 days of data for Baltimore County.

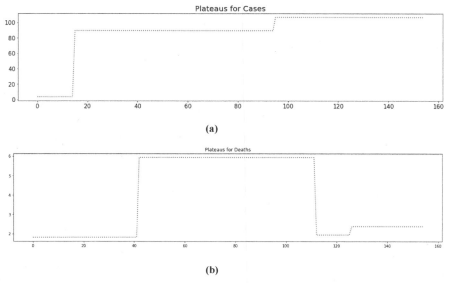

Figure 7.12 COVID-19 distributions for Baltimore County cases and deaths. (a) Distributions of COVID-19 cases (*initial number of bins: 155; number of partitioned bins: 40; distance measure: Mahalanobis; threshold: 0.15; number of merged bins: 3*). (b) Distributions of COVID-19 deaths (*initial number of bins: 155; number of partitioned bins: 25; distance measure: Mahalanobis; threshold: 0.15; number of merged bins: 4*).

For the distributions of cases, namely Figure 7.13(a), as already mentioned, we had 155 days of data which corresponded to 155 initial bins for the SMerg algorithm; these bins were merged into 75 and 45 bins. Mahalanobis distance metric was used and the threshold was set to 0.65. Mean of these 45 discretized bins represent the 45 temporal neighborhoods for the COVID-19 cases in Baltimore County. Further, in order to generate distributions of deaths for Baltimore County as shown in Figure 7.13(b), initial bins were 155; these bins were partitioned into 49 and 20 bins with the varied parameter settings.

As we can see, Figure 7.12 aligns with the overall distribution of the raw data. When we vary the parameters such as partitioned bins, distance measure, and threshold, it is possible to obtain a higher number of merged bins. Mean distribution of the values obtained from the merged bins helps in visualizing the trend of COVID-19 cases and deaths.

We also included New York City (NYC) in our analysis as it was also severely impacted by COVID-19. In terms of the city dynamics, it was a comparison point to Baltimore City. However, it also stood out as a very

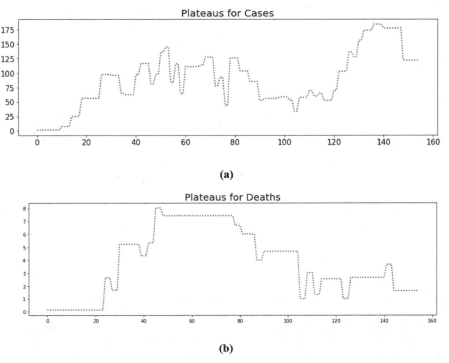

(a)

(b)

Figure 7.13   COVID-19 distributions for Baltimore County cases and deaths using different combinations of parameters for SMerg algorithm. (a) Distributions of COVID-19 cases (*initial number of bins: 155; number of partitioned bins: 75; distance measure: Mahalanobis; threshold: 0.65; number of merged bins: 45*). (b) Distributions of COVID-19 deaths (*initial number of bins: 155; number of partitioned bins: 49; distance measure: Mahalanobis; threshold: 0.33; number of merged bins: 20*).

different feature set as compared to our other locations. Figure 7.14 shows the COVID-19 distributions generated for New York City. It can be observed that between the 10th day and the 80th day, the number of cases exploded. We also observed that the cases in NYC were on a very different scale as compared to others. This aligns with the nature of things during the outbreak in NYC as compared to other locations. This could also be due to the high density in NYC along with the large number of assisted living facilities [18]. The number of cases reported were higher than 1000 on a daily basis. The number of cases has now significantly reduced as compared to the beginning. Figures 7.11–7.13 show distribution of cases across Montgomery and Baltimore County. While the distributions look similar, the strength of the spread in terms of number of cases is still quite different.

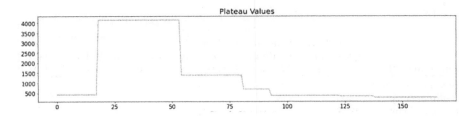

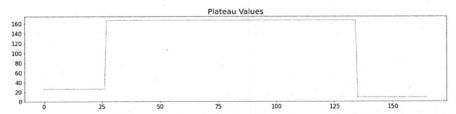

Figure 7.14 COVID-19 distributions for New York City cases and deaths. (a) Distributions of COVID-19 cases (*initial number of bins: 166; number of partitioned bins: 50; distance measure: Mahalanobis; threshold: 0.03; number of merged bins: 7*). (b) Distributions of COVID-19 deaths (*initial number of bins: 166; number of partitioned bins: 50; distance measure: Mahalanobis; threshold: 0.03; number of merged bins: 3*).

Figures 7.14(a) and (b) both depict the distributions of COVID-19 cases and deaths reported in New York City. 166 days of data is collected for New York City as the first COVID-19 case in New York City was reported earlier than that in Maryland. First case in New York City was reported on 29 February 2020. Using SMerg algorithm, we were able to generate robust distributions by computing the mean of the distributions obtained from the merged bins.

It can be observed that the number of steps in each distribution corresponds to the number of merged bins obtained from the SMerg algorithm.

Table 7.2. shows summarization of trends for number of cases and deaths across different locations impacted by COVID-19.

Table 7.2 helps in analyzing the spread of COVID-19 across five different counties of the state of Maryland and New York City. For New York City, the trend remains high for a major portion of the analysis. This validates that New York City was severely affected by COVID-19. In the state of Maryland, Baltimore County and Prince George's county have been impacted by COVID-19 related deaths. Howard County has the least number of deaths reported as compared to other locations. The distributions of New York City, Prince George's County, and, to some extent, Montgomery County and Baltimore County reveal a similar

Table 7.2   Summarization of trends for number of cases and deaths across different locations impacted by COVID-19.

Location Name	Cumulative Cases till 12th August	Cumulative Death till 12th August	Current positivity rate	Cases trend till 12th Aug	Death trend till 12th Aug
Baltimore Country	13485	584	8.40%		
Baltimore city	13007	434	4.30%		
Howard Country	3947	112	4.80%		
Montgomery Country	18650	806	5.60%		
Prince George's Country	24560	764	10.00%		
New York City	233216	19078	1%		

trend of larger numbers of deaths from the distribution. The other locations are on a gradual decline. The distribution of NYC, Montgomery County, Prince George's County, and Baltimore County can be used to search other locations to identify similar behavior in terms of the pandemic spread.

7.4 Conclusion

In this chapter, we have described an approach to find anomalies and distributions of temporal data with similar and dissimilar patterns in terms of spread of COVID-19. We have largely assessed the COVID-19 situation via number of confirmed cases of COVID-19 and number of deaths reported due to COVID-19. Distributions generated using temporal neighborhood approach for each county's visualization shows the trend of the spread of COVID-19. Distributions also facilitate an analysis to distinguish when there was a trend in cases and deaths reported. Through our experiments, we believe that we can accelerate the research efforts to understand the distribution of COVID-19 across different geospatial locations and compare the impact of policies implemented. We further intend to extend our study to analyze the proliferation of COVID-19 across similar and dissimilar demographics. We intend to extend this study by generating samples of distributions associated with geospatial features so that they can be searchable to provide insights from similar types of spreads.

References

[1] Dey, Sandipan, Vandana P. Janeja, and Aryya Gangopadhyay. "Temporal neighborhood discovery using markov models." *2009 Ninth IEEE International Conference on Data Mining*. IEEE, 2009.

[2] Yeh, Chin-Chia Michael, et al. "Matrix profile I: all pairs similarity joins for time series: a unifying view that includes motifs, discords and shapelets." *2016 IEEE 16th international conference on data mining (ICDM)*. Ieee, 2016.

[3] Guo, Chonghui, Hailin Li, and Donghua Pan. "An improved piecewise aggregate approximation based on statistical features for time series mining." *International conference on knowledge science, engineering and management*. Springer, Berlin, Heidelberg, 2010.

[4] Cheng, Zhangyu, Chengming Zou, and Jianwei Dong. "Outlier detection using isolation forest and local outlier factor." *Proceedings of the conference on research in adaptive and convergent systems*. 2019.

[5] Basu, Sabyasachi, and Martin Meckesheimer. "Automatic outlier detection for time series: an application to sensor data." *Knowledge and Information Systems* 11.2 (2007): 137-154.

[6] Fu, Tak-chung. "A review on time series data mining." *Engineering Applications of Artificial Intelligence* 24.1 (2011): 164-181.

[7] Janeja, Vandana P., et al. "Spatial neighborhood based anomaly detection in sensor datasets." *Data Mining and Knowledge Discovery* 20.2 (2010): 221-258.

[8] Kullback, Solomon, and Richard A. Leibler. "On information and sufficiency." *The annals of mathematical statistics* 22.1 (1951): 79-86.

[9] Mohammadi, Seyed H., Vandana P. Janeja, and Aryya Gangopadhyay. "Discretized spatio-temporal scan window." *Proceedings of the 2009 SIAM International Conference on Data Mining*. Society for Industrial and Applied Mathematics, 2009.

[10] Panconesi, Alessandro. "The stationary distribution of a Markov chain." *Unpublished note, Sapienza University of Rome, http://www. dis. uniroma1. it/~ leon/didattica/webir/pagerank. pdf* (2005).

[11] Kriegel, Hans-Peter, Peer Kröger, and Arthur Zimek. "Outlier detection techniques." Tutorial at KDD 10 (2010): 1-76.

[12] Smith, Mitch, et al. "Tracking Every Coronavirus Case in the U.S.: Full Map." *The New York Times*, 2020, www.nytimes.com/interactive/2020/us/coronavirus-us-cases.html.

[13] Hogan, Larry, et al. State Plan on Aging. 201, URL: https://aging.maryland.gov/Documents/MDStatePlan2017_2020Dated092216.pdf Accessed 16 Sept. 2020.

[14] Anomaly Detection Using Isolation Forest in Python. *Paperspace Blog*, 2 Mar. 2020, blog.paperspace.com/anomaly-detection-isolation-forest/. Accessed 16 Sept. 2020.

[15] StoryMapSeries.Princegeorges.Maps.Arcgis.Com,princegeorges.maps. arcgis.com/apps/MapSeries/index.html?appid= 82fa5c47b1f542849ca6162ab1564453. Accessed 16 Sept. 2020.

[16] COVID-19 Pandemic in Maryland. Wikipedia, 14 Sept. 2020, en.wikipedia.org/wiki/COVID-19_pandemic_in_Maryland. Accessed 16 Sept. 2020.

[17] U.S. Census Bureau QuickFacts: Prince George's County, Maryland. Www.Census.Gov, www.census.gov/quickfacts/fact/table/ princegeorgescountymaryland/PST045219. Accessed 16 Sept. 2020.

[18] NYS Adult Care Facility Profiles We make it easy to find quality and safety information on New York's adult care facilities.URL:https:// profiles.health.ny.gov/acf Website Title: New York State Department of Health Date Accessed:September 17, 2020

[19] Coronavirus In Maryland: Cases Top 46K; Hospitalizations Drop Below 1,300 URL:https://baltimore.cbslocal.com/2020/05/24/coronavirus-in-maryland-46k-cases-hospitalizations-below-1300/ Website Title:CBS Baltimore Date Accessed:September 17, 2020

[20] Baltimore County sees increase in positive COVID-19 cases, releases new health guidelines URL:https://www.avenuenews.com/news/ local/baltimore-county-sees-increase-in-positive-covid-19-cases-releases-new-health-guidelines/article_21f4faaa-ecd5-5e31-8b96-e45a3e86f392.html Website Title:The Avenue News Date Accessed: September 17, 2020

[21] Coronavirus In Maryland: Total Cases Climb More Than 900 To 81.7K; Hospitalizations Nearly Flat URL: https://baltimore.cbslocal. com/2020/07/24/coronavirus-in-maryland-total-cases-nearly-82k-hospitalizations-flat/ Website Title: CBS Baltimore Date Accessed: September 17, 2020

[22] COVID-19: Data URL: https://www1.nyc.gov/site/doh/covid/covid-19-data.page Website Title: COVID-19: Data Main - NYC Health Date Accessed: September 17, 2020

[23] Maryland Department of Health URL:https://coronavirus.maryland. gov/ Website Title: Coronavirus Date Accessed: September 17, 2020

[24] Anomaly Detection: Matrix Profile Discords URL:https://tylermarrs. com/posts/anomaly-detection-matrix-profile-discords/ Website title:Tyler Marrs Date Accessed:September 18, 2020

[25] Piecewise Aggregate Approximation URL: https://johannfaouzi. github.io/pyts/auto_examples/plot_paa.html Piecewise Aggregate Approximation - pyts 0.7.0 documentation Date Accessed:September 18, 2020

8

Toward Smart Hospital: An Intelligent Personnel Scheduling Using Evolutionary Algorithms

Tan Nhat Pham[1] **and Son Vu Truong Dao**[1]

[1]International University, Vietnam National University, Ho Chi Minh City, Vietnam
Corresponding Author: Son Vu Truong Dao, dvtson@hcmiu.edu.vn

Abstract

The nurse scheduling problem (NSP) is defined as the operation research of assigning shifts to available nurses over a planning period with a number of constraints. We propose a method by grouping nurses into variable clusters, in which each cluster is served by a schedule improved by a grey wolf optimization (GWO). We obtain the results from GWO and compare with those from particle swarm optimization (PSO) and IBM CPLEX Studio Solver to evaluate the effectiveness of our algorithm.

8.1 Introduction

The main purpose of all nurse scheduling problems (NSPs) is to create the schedules that both satisfy the fit constraints for nurses and the purposes that the hospital would like to achieve. There are three shifts for a nurse to work: morning, evening, night shift, and day-off. For circumstances happening in all corner of the world today, the COVID-19 pandemic, almost all hospitals have faced the huge shortage of nurses and doctors, which leads to the fact

111

that they have to stay at hospitals without coming back home for over a month. This paper hopes to solve that problem once it happens again.

The NSP or nurse rostering problem (NRP) [1] is the operation of finding the best way to assign attendants to working shifts, regularly with a set of difficult imperatives which all substantial arrangements must take after, and a set of delicate limitations which characterize the quality of substantial arrangements. Other fields can utilize the solutions to NSP as well. Before 1969, this type of problem has been studied and is known to have an NP-hard difficulty [2].

NSP are previously solved using linear [3] or quadratic programming [4] depending on the constraints. They may also be formulated and solved using search methods, including tabu search [5].

Grey wolf optimizer (GWO) [6] is a metaheuristic to solve problems using hunting behaviors of wolves. It had shown good results on many fields, such as 3D Stacked System-on-Schip (SoC) [7], Job Shop [8], Clustering wireless sensor network (WSN) [9], and Prediction [10].

In this study, we apply a GWO to solve NSPs by minimizing a number of nurses needed for allocation of shift. We also compare the results with those having particle swarm optimization (PSO) and exact solution. PSO was also successfully implemented to solve many combinatorial optimization problems [11–13]. The authors in [14] conducted a review on prominent nature-inspired algorithm, including their pros and cons, in which GWO outperformed other algorithms on benchmark functions. Searching mechanisms of GWO and PSO are illustrated in Figure 8.1.

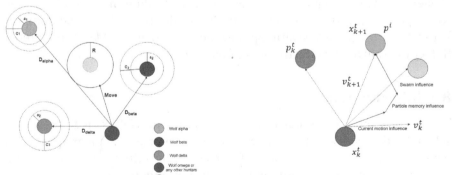

Figure 8.1   Searching mechanism of GWO (left) and PSO (right).

8.2 Methodology

8.2.1 Data Collection

This paper uses the datasets which are being reputable and public internationally, in terms of 24 instances, obtained from [15]. By challenging collection of benchmark test instances selected from several sources, we can test and develop the GWO algorithm, especially in some fields of industrial collaborators and scientific publications, which is the main purpose of these datasets.

8.2.2 Mathematical Model Development

To solve NSP, satisfying the demand is the most important thing, the other things like consecutive shifts and days or minimum and maximum working days.

One day has three shifts: morning (7.00–16.00), evening (16.00–23.00), and night (23.00–7.00).

Index:

i: index of nurse $i = 1, ..., I$

d: index of day $d = 1, ..., D$

D: number of days in the scheduling period

I: number of nurse

Parameters:

M_d: number of nurses required for morning shift d

E_d: number of nurses required for evening shift d

N_d: number of nurses required for night shift d

Decision variables:

XM_{id} binary variable $XM_{id} = 1$ if nurse i is assigned to morning shift d; otherwise, $XM_{id} = 0$.

XE_{id} binary variable $XE_{id} = 1$ if nurse i is assigned to evening shift d; otherwise, $XE_{id} = 0$.

XN_{id} binary variable $XN_{id} = 1$ if nurse i is assigned to night shift d; otherwise, $XN_{id} = 0$.

XR_{id} binary variable $XR_{id} = 1$ if nurse i is off on day d, and 0 otherwise.

$Y_i = 1$ if nurse i is used, and 0 otherwise.

C_i is the cost for nurse i, \$100, \$150, \$200, and \$0 for morning, evening, night, and day-off, respectively.

Objective function: Minimize the cost based on the number of nurses needed

$$\textbf{Minimize} \sum\nolimits_{i=1}^{I} C_i Y_i \tag{8.1}$$

Constraint:

Demand constraints: Each shift must be enough for the minimum coverage demand of nurses

$$\sum_{i=1}^{I} XE_{id} \geq E_d, \forall d \tag{8.2}$$

$$\sum_{i=1}^{I} XM_{id} \geq M_d, \forall d \tag{8.3}$$

$$\sum_{i=1}^{I} XN_{id} \geq N_d, \forall d \tag{8.4}$$

The consecutive shift constraints:

One nurse cannot work two consecutive shifts and at most two shifts a day.

A nurse working at night shift cannot work in the morning shift.

$$XE_{id} + XM_{i,d+1} = 1 \tag{8.5)}$$

$$XE_{id} + XN_{id} = 1 \tag{8.6)}$$

$$XN_{id} + XM_{i,d+1} = 1 \tag{8.7)}$$

One nurse cannot work at most six consecutive days and five shifts per week.

$$\sum_{t=0}^{t=6} XR_{i,d+t} \geq 1, \forall i, d = 1, \ldots, D - 6 \tag{8.8}$$

$$\sum_{t=0}^{t=6} XM_{i,d+t} + \sum_{t=0}^{t=6} XE_{i,d+t} + \sum_{t=0}^{t=6} XN_{i,d+t} \leq 5, \forall i, d = 1,8,15,22,\ldots \tag{8.9}$$

If the nurse i is used, then $Y_i = 1$

$$Y_i \geq XM_{id}, \forall i,d \tag{8.10}$$

$$Y_i \geq XE_{id}, \forall i,d \tag{8.11}$$

$$Y_i \geq XN_{id}, \forall i,d \tag{8.12}$$

$$Y_i \geq \sum_{d=1}^{D} XM_{id} + \sum_{d=1}^{D} XE_{id} + \sum_{d=1}^{D} XN_{id}, \quad \forall i \tag{8.13}$$

8.2.3 Discrete GWO with a Novel Neighborhood Search Operator

In order to apply GWO to NSP, we define the following terminology:

- Search agent/wolf: a random schedule of all nurses.
- Prey: the minimum cost of all schedules.
- Fitness: total cost of a schedule with all the constraints.
- Local best: minimum fitness of alpha, beta, and delta wolves in the local population.
- Global best: minimum fitness of alpha, beta, and delta wolves in the global population.
- Best fitness: fitness value that can be reached after a certain number of iterations when GWO has optimized the schedule that satisfies all constraints.

8.2.3.1 Population Initialization

The assignment of nurses is randomly generated; each random schedule represents one search agent. The search space is a matrix with size $N \times D$, in which N is the number of search agents and D is the schedule period, measured in days. The three best schedules having three best fitness values are defined as wolf α, β, and δ, whose positions are X_α, X_β, and X_δ. The shifts are denoted by number: number 1 stands for the morning shift (M) and numbers 2, 3, and 4 represent the evening shift (E), night shift (N), and day-off (O), respectively. If there are 10 nurses, N and the schedule period is 14 days, then the search space matrix would have $10 \times 14 = 140$ elements. Elements 1–14 represent the schedule of nurse 1, 15–28 for nurse 2, 29–42 for nurse 3, and so on. This encoding process is shown in Table 8.1.

Table 8.1   A sample matrix of initial position for GWO.

Grey wolf	(1,1)	...	(1,14)	(2,15)	...	(2,28)	...	(N,N*D)
Wolf 1	M	...	N	N	...	E	...	M
.	A	...	M	O	...	E	...	N
Wolf 2	O	...	M	E	...	M	...	N
.	N	...	E	O	...	N	...	E
Wolf i	N	...	M	N	...	E	...	O
.	O	...	E	N	...	O	...	E
Wolf N	E	...	N	E	...	O	...	M

8.2.3.2 Discrete Search Strategy

The overall d GWO is originally designed for continuous optimization problems. In this work, we employed a discrete operator based on the crossover and mutation inspired from genetic algorithm. Let X_k^t be the solution of the kth wolf; X_α^t, X_β^t, and X_δ^t define the solutions of wolf α, β, and δ at iteration t; and let be the crossover operation, and be a random number in range [0,1]. Then we have the following position updating process:

$$X_k^{t+1} = \begin{cases} f\left(X_k^t, X_\alpha^t\right), & \text{if } \text{rand} \le \dfrac{1}{3} \\ f\left(X_k^t, X_\beta^t\right), & \text{if } \dfrac{1}{3} \text{ rand} \le \dfrac{2}{3} \\ f\left(X_k^t, X_\delta^t\right), & \text{if } \text{rand} \le \dfrac{2}{3} \end{cases} \tag{8.14}$$

The procedure of the crossover operation is as follows:
- Divide search agents by a half into two subsets.
- Randomly select two search agents to be the parents.
- Choose a random crossover-point.
- Copy the first part of Parent 1 and merge with the second part of Parent 2 to create new solution, called Child 1, and vice versa to create Child 2.

In this work, we implemented both one- and two-point crossover operators. The mechanism is shown in Figures 8.2 and 8.3. During the main loop of the algorithm, each operator is chosen with a probability of 0.5 in order to enhance the diversity of the population.

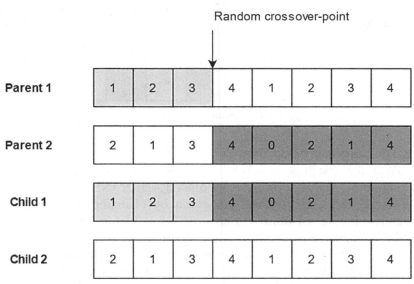

Figure 8.2   One-point crossover operation.

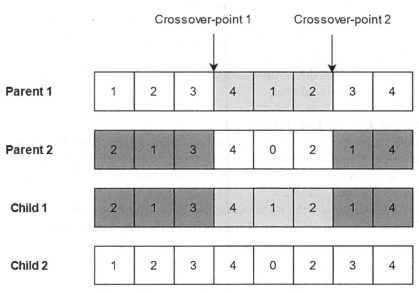

Figure 8.3   Two-point crossover operation.

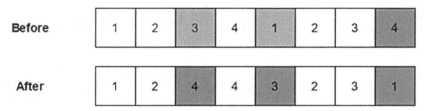

Figure 8.4   Creating neighborhood solution.

8.2.3.3 Neighborhood Search

In GWO, the search process is governed by three best (α, β, and δ) toward the optimum. We perform a neighborhood search (NS) on the three best individuals. NS can improve the search by changing the neighborhood structures. Figure 8.4 illustrates how the NS is performed. First, a random number n is generated in between 1 and the length of the solution. Then n points would be randomly chosen to be swapped by picking n random numbers R_1, $R_2,\ldots, R_n$ which lie between 0 and the total elements in the solution list after defining a function initialization with input being the number of solution. If any two of them are equal, then they would be re-generated. The next step is creating a blank list named X_{temp}. In the solution array, let A_1 be the element in the index R_1, A_2 be the element with index R_2, and similarly up to A_n. The elements are then swapped in a complete random manner and added to X_{temp} list, and then all the remaining elements of the original solution are also added to the new X_{temp}. The NS operators are also used during the main loop to further increase the diversity, triggered by a probability $P_n = 0.3$; this value is found by empirical study.

8.2.3.4. Constraint Handling

All the solutions' fitness values are calculated using the objective function and violation of constraints is also being checked. If there are violated constraints, a penalty value is added to the objective function so that the algorithm will discard the bad solution.

Here we present the pseudocode for the algorithm in Figure 8.5. With the above design, the procedure of the GWO is illustrated in Figure 8.6.

Initialize the population of the wolves

Intitialize parameters

Calculate the fitness value of all wolves

Alpha = Best wolf

Beta = Second best wolf

Delta = Third best wolf

while *(t< Max number of iteration)*

    for *each search agent*

        Update the positions by the discrete strategy

        Perform NS with P_n

    endfor

    Calculate fitness of all agents

    Perform constraint handling for agents' fitness

    Update Alpha, Beta, Delta

    Perform NS for Alpha, Beta, Delta

    Update Alpha, Beta, Delta again

    t = t+1

end while

return *Alpha and Alpha's fitness*

Figure 8.5 Pseudocode of the solution.

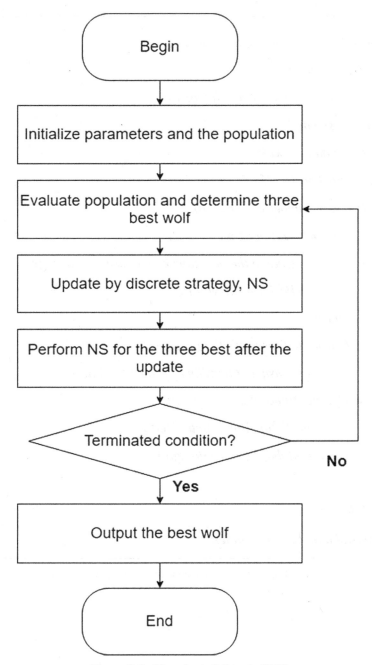

Figure 8.6   Flowchart of discrete GWO.

8.3 Computational Results

In this part, the solutions found by GWO are compared with ones found by IBM CPLEX and PSO. Demand of each instance is written in the form of (x, y, z), which means it requires x nurses for the morning shift, y nurses for the evening shift, and z nurses for the night shift. Both CPLEX and Python program are run on a PC with an Intel Core i7 CPU at 3.40 GHz with 16 GB of RAM. The average result of five consecutive runs by GWO and PSO are recorded in Table 8.2.

Table 8.2   Summarized result from 33 datasets.

Instance	Demand	Nurse	Day	Cost by GWO	Cost by PSO	Cost by CPLEX
1	(5,0,0)	8	14	8350	8550	7000
2	(0,5,4)	14	14	21,850	22,350	21,700
3	(3,5,3)	20	14	27,050	29,100	23,100
4	(3,0,3)	10	28	28,650	25,900	25,200
5	(6,0,4)	16	28	39,250	42,300	392,00
6	(4,3,4)	18	28	47,050	48,250	46,200
7	(4,5,2)	20	28	48,900	54,000	43,400
8	(4,4,3)	30	28	76,650	85,900	44,800
9	(4,4,3)	36	28	51,650	104,300	44,800
10	(5,6,2)	40	28	65,600	121,000	50,400
11	(9,11,9)	50	28	132,550	148,000	121,800
12	(10,11,11)	60	28	166,400	178,200	135,800
13	(17,17,18)	120	28	338,150	365,700	219,800
14	(5,5,4)	32	42	126,450	150,150	86,100
15	(8,5,5)	45	42	187,950	201,700	107,100
16	(3,5,3)	20	56	100,750	112,600	92,400
17	(5,6,5)	32	32	93,900	115,400	76,800
18	(3,5,3)	22	48	94,850	106,750	79,200
19	(7,6,7)	40	48	186,500	192,400	144,000
20	(5,15,5)	50	104	536,550	574,000	390,000
21	(5,5,6)	100	104	790,650	803,650	655,200
22	(3,3,3)	50	208	1,135,500	1,820,700	1,123,200
23	(11,14,11)	100	208	1,155,000	1,215,500	1,123,200
24	(15,19,14)	150	208	1,523,100	1,809,300	1,487,200
25	(3,3,3)	24	14	18,950	19,450	18,900
26	(4,3,4)	28	14	23,250	23,500	23,100
27	(4,4,4)	28	14	25,400	25,650	25,200
28	(1,1,1)	10	14	8050	27,450	6300
29	(2,1,2)	10	14	10,600	25,150	10,500
30	(3,1,3)	20	14	20,300	27,150	14,700

Continued

Table 8.2   Continued

Instance	Demand	Nurse	Day	Cost by GWO	Cost by PSO	Cost by CPLEX
31	(3,2,3)	20	14	21,750	27,200	16,800
32	(11,20,11)	100	104	1,103,550	Infeasible	Infeasible
33	(8,8,8)	50	208	1,251,400	Infeasible	Infeasible
34	(9,9,9)	50	14	135,450	Infeasible	Infeasible
35	(10,10,10)	50	14	156,500	Infeasible	Infeasible
36	(11,11,11)	40	14	166,450	Infeasible	Infeasible
37	(12,12,12)	80	14	195,500	Infeasible	Infeasible
38	(10,11,12)	80	14	210,500	Infeasible	Infeasible
39	(13,13,13)	80	14	245,650	Infeasible	Infeasible
40	(13,13,13)	80	14	315,450	Infeasible	Infeasible

The gap from the GWO and PSO results with respect to exact solution is calculated using the following equation:

$$\frac{\text{Results obtained by PSO} - \text{Results obtained by CPLEX}}{\text{Results obtained by CPLEX}} \qquad (8.15)$$

$$\frac{\text{Results obtained by GWO} - \text{Results obtained by CPLEX}}{\text{Result obtained by CPLEX}} \qquad (8.16)$$

The gap is presented in Table 8.3.

Table 8.3   Summarized gap from 33 datasets.

Instance	Demand	Nurse	Day	Gap (GWO-CPLEX) (%)	Gap (PSO-CPLEX) (%)
1	(5,0,0)	8	14	19.29	22.14
2	(0,5,4)	14	14	0.69	3.00
3	(3,5,3)	20	14	17.10	25.97
4	(3,0,3)	10	28	13.69	2.78
5	(6,0,4)	16	28	0.13	7.91
6	(4,3,4)	18	28	1.84	4.44
7	(4,5,2)	20	28	12.67	24.42
8	(4,4,3)	30	28	71.09	91.74
9	(4,4,3)	36	28	15.29	132.81
10	(5,6,2)	40	28	30.16	140.08
11	(9,11,9)	50	28	8.83	21.51
12	(10,11,11)	60	28	22.53	31.22
13	(17,17,18)	120	28	53.84	66.38
14	(5,5,4)	32	42	46.86	74.39

Continued

Table 8.3 Continued

Instance	Demand	Nurse	Day	Gap (GWO-CPLEX) (%)	Gap (PSO-CPLEX) (%)
15	(8,5,5)	45	42	75.49	88.33
16	(3,5,3)	20	56	9.04	21.86
17	(5,6,5)	32	32	22.27	50.26
18	(3,5,3)	22	48	19.76	34.79
19	(7,6,7)	40	48	29.51	33.61
20	(5,15,5)	50	104	37.58	47.18
21	(5,5,6)	100	104	20.67	22.66
22	(3,3,3)	50	208	1.10	62.10
23	(11,14,11)	100	208	2.83	8.22
24	(15,19,14)	150	208	2.41	21.66
25	(3,3,3)	24	14	0.26	2.91
26	(4,3,4)	28	14	0.65	1.73
27	(4,4,4)	28	14	0.79	1.79
28	(1,1,1)	10	14	27.78	335.71
29	(2,1,2)	10	14	0.95	139.52
30	(3,1,3)	20	14	38.10	84.69
31	(3,2,3)	20	14	29.46	61.90
32	(11,20,11)	100	104	--	--
33	(8,8,8)	50	208	--	--
34	(9,9,9)	50	14	--	--
35	(10,10,10)	50	14	--	--
36	(11,11,11)	40	14	--	--
37	(12,12,12)	80	14	--	--
38	(10,11,12)	80	14	--	--
39	(13,13,13)	80	14	--	--
40	(13,13,13)	80	14	--	--

8.4 Conclusion

The solutions show that GWO outperformed PSO in all instances, although for complicated cases, the solution found by GWO is still far from optimum. One point worth noting is that for large instances such as instances 32 and 33, PSO and CPLEX cannot find feasible solutions. On average, GWO's solution is approximately 20.4% larger than the exact solution, while the mean gap of solution found by PSO is about 53.8%. For many instances, such as instance 2, 5, 26, or 27, GWO provides solutions that are very close to the exact solution. It can be clearly seen that GWO is more stable and better at avoiding local optima, at which PSO usually cannot escape from. Besides saving the budget of the hospital, this can allocate the good schedules in a short time rather than manually assigning shifts to each nurse as a lot of hospitals have been doing right now. Besides, PSO can only solve the small case in suitable time; however, with the instance with large number of nurses and schedule period, it took too much time to run the same number of iterations as GWO. Also, there were still complicated instances that GWO outperformed the exact method. Therefore, this work is considered helpful in the future research of GWO as well as health care system. This work proposed a new metaheuristic approach to solve NSP which has never been used in this field. The results demonstrate that GWO gives much better total cost compared to that using PSO. Furthermore, results obtained using GWO are relatively close to the exact solution using CPLEX for small case study. However, for large case studies, due to combinatorial nature of NSP, there is no solution available using the exact method and our GWO provides the best results. In future works, we would like to extend our work by working with other scheduling scenarios with more practical constraints and involved other variants of GWO. One of the potential candidates is the hybrid adaptive PSO and GWO proposed by [16].

References

[1] A. T. Ernst, H. Jiang, M. Krishnamoorthy, and D. Sier, "Staff scheduling and rostering: A review of applications, methods and models," *Eur. J. Oper. Res.*, vol. 153, no. 1, pp. 3–27, Feb. 2004, doi: 10.1016/S0377-2217(03)00095-X.

[2] I. Solos, I. Tassopoulos, and G. Beligiannis, "A Generic Two-Phase Stochastic Variable Neighborhood Approach for Effectively Solving the Nurse Rostering Problem," *Algorithms*, vol. 6, no. 2, pp. 278–308, May 2013, doi: 10.3390/a6020278.

[3] L. Trilling, A. Guinet, and D. Le Magny, "NURSE SCHEDULING USING INTEGER LINEAR PROGRAMMING AND CONSTRAINT PROGRAMMING," *IFAC Proc. Vol.*, vol. 39, no. 3, pp. 671–676, Jan. 2006, doi: 10.3182/20060517-3-FR-2903.00340.

[4] H. E. Miller, W. P. Pierskalla, and G. J. Rath, "NURSE SCHEDULING USING MATHEMATICAL PROGRAMMING.," *Oper. Res.*, vol. 24, no. 5, pp. 857–870, Oct. 1976, doi: 10.1287/opre.24.5.857.

[5] K. A. Dowsland, "Nurse scheduling with tabu search and strategic oscillation," *Eur. J. Oper. Res.*, vol. 106, no. 2–3, pp. 393–407, Apr. 1998, doi: 10.1016/S0377-2217(97)00281-6.

[6] S. Mirjalili, S. M. Mirjalili, and A. Lewis, "Grey Wolf Optimizer," *Adv. Eng. Softw.*, vol. 69, pp. 46–61, 2014, doi: 10.1016/j.advengsoft.2013.12.007.

[7] A. Zhu, C. Xu, Z. Li, J. Wu, and Z. Liu, "Hybridizing grey Wolf optimization with differential evolution for global optimization and test scheduling for 3D stacked SoC," *Journal of Systems Engineering and Electronics*, vol. 26, no. 2. pp. 317–328, 2015, doi: 10.1109/JSEE.2015.00037.

[8] T. Jiang and C. Zhang, "Application of Grey Wolf Optimization for Solving Combinatorial Problems: Job Shop and Flexible Job Shop Scheduling Cases," *IEEE Access*, vol. 6, pp. 26231–26240, 2018, doi: 10.1109/ACCESS.2018.2833552.

[9] M. Sharawi and E. Emary, "Impact of grey Wolf optimization on WSN cluster formation and lifetime expansion," in *9th International Conference on Advanced Computational Intelligence, ICACI 2017*, Jul. 2017, pp. 157–162, doi: 10.1109/ICACI.2017.7974501.

[10] M. Wang *et al.*, "Grey wolf optimization evolving kernel extreme learning machine: Application to bankruptcy prediction," *Eng. Appl. Artif. Intell.*, vol. 63, pp. 54–68, Aug. 2017, doi: 10.1016/j.engappai.2017.05.003.

[11] T. H. Wu, J. Y. Yeh, and Y. M. Lee, "A particle swarm optimization approach with refinement procedure for nurse rostering problem," *Comput. Oper. Res.*, vol. 54, pp. 52–63, Feb. 2015, doi: 10.1016/j.cor.2014.08.016.

[12] M. A. Hannan, M. Akhtar, R. A. Begum, H. Basri, A. Hussain, and E. Scavino, "Capacitated vehicle-routing problem model for scheduled solid waste collection and route optimization using PSO algorithm," *Waste Manag.*, vol. 71, pp. 31–41, Jan. 2018, doi: 10.1016/j.wasman.2017.10.019.

[13] M. Eddaly, B. Jarboui, and P. Siarry, "Combinatorial particle swarm optimization for solving blocking flowshop scheduling problem," *J. Comput. Des. Eng.*, vol. 3, no. 4, pp. 295–311, Oct. 2016, doi: 10.1016/j.jcde.2016.05.001.

[14] M. J. Islam, M. S. R. Tanveer, and M. A. H. Akhand, "A comparative study on prominent nature inspired algorithms for function optimization," in *2016 5th International Conference on Informatics, Electronics and Vision, ICIEV 2016*, Nov. 2016, pp. 803–808, doi: 10.1109/ICIEV.2016.7760112.

[15] T. Curtois and R. Qu, "Computational results on new staff scheduling benchmark instances," 2014. Accessed: Aug. 20, 2020. [Online]. Available: http://schedulingbenchmarks.org/papers/computational_results_on_new_staff_scheduling_benchmark_instances.pdf.

[16] D. Son and P. N. Tan, "Capacitated Vehicle Routing Problem - A New Clustering Approach Based on Hybridization of Adaptive Particle Swarm Optimization and Grey Wolf Optimization," in *Evolutionary Data Clustering: Algorithms, and Applications*, Springer, 2019.

9

Role of Artificial Intelligence Based Wireless Sensor Network for Pandemic Control: A Case Study Using CupCarbon

Paawan Sharma, Hardik Patel, and Mohendra Roy

Department of Information and Communication Technology, Pandit Deendayal Petroleum University, Gandhinagar, Gujarat 382007, India
Corresponding Author: Paawan Sharma, paawan.sharma@sot.pdpu.ac.in

Abstract

Occurrence and spread of COVID-19 has posed numerous difficult challenges for the world. At the same time, it has also provided an opportunity for the researchers in various domains to adopt a sustainable approach in fighting the challenge of COVID-19. The present work reports a case study on the use of artificial intelligence based wireless sensor network for pandemic control. The proposed system is simulated in CupCarbon, which is a smart city simulator for different scenarios. The analysis presented in the paper helps to understand the impact of using network algorithms, complexity, and size of the network in mitigating the pandemic spread. Also, it helps the smart city designers to accommodate the tools required for such damage control mechanisms.

9.1 Introduction

Role of WSN and IoT in disaster management: Wireless sensor network (WSN) and the Internet of Things (IoT) can provide state-of-the-art solutions for many disaster scenarios [1]. Together with advanced artificial intelligence

(AI) algorithms, we can design a smart disaster management system. Natural hazards and calamities are the major contributors to disaster. That is why it is in priority in disaster management to monitor and predict any such calamity. This is also a very challenging issue. In this regard, IoT powered WSN could be a good solution. SyedIjlal *et al.* have described the use of WSN in flood-related natural hazards [2]. Their WSN Aqua-net is a robust network to monitor water management systems even in flood-like situations.

Again, the recent wildfires in Australia and India have shown us how devastating the wildfire can be. However, there is very little arrangement for this kind of scenario that can provide an alert. Hsu-Yang *et al.* have developed a WSN to mitigate this problem [3]. They have developed and deployed WSN for drought monitoring which, in turn, provides possible wildfire detection also. Further, they have incorporated a neural network to analyze the data from the sensor network and predict the behavior of the environment. They have also developed an emergency action inference engine (EAIE) to provide inference about the measures that need to be placed. Landslides, especially in the hilly area, are also a major contributor to disaster. In this regard, the WSN of vibration sensors may help to provide early warning. In this regard, Rossi *et al.* have described the real-world applications by deploying WSN sensors [4]. Ramesh *et al.* showed a real-world model of WSN for landslide detection which shows its efficiency at the time of the flood in 2009 in Kerala [5].

Role of WSN and IoT for smart cities: For a smart city, it is essential that all its fundamental components have the basic characteristics of a smart city. These fundamental components are smart building, smart transport, and smart traffic management. However, for a smart city, all these components should be connected. In this regard, WSN with IoT can play a major role. The following are the current developments in these areas.

WSN in smart buildings: Connection and automation in buildings makes a building smart. Mingze *et al.* summarized ZigBee WSN and its applications in smart home and building automation [6]. Again, Minoli *et al.* optimize the IoT requirements and their architectures for smart building applications [7]. These optimized IoT networks can also be utilized for the safety of the residents of the buildings by providing alarm to the central stations in case of emergency, such as fire. Islam *et al.* summarize a fire detection system based on ZigBee network for smart buildings.

Climate change is a very serious issue for the whole world. It results in natural disasters like storms, floods, drought, and heat waves in different parts of the world. Rapid industrialization and urbanization are creating carbon emission or global warming, noise, dust, and temperature rise problems. Monitoring system is designed and implemented for early detection and warning generation about the natural disasters and environmental threats in Quang Nam and Da Nang provinces in central Vietnam [8]. The integrated WSN system provides temperature, CO_2 levels, dust, and noise in Da Nang. It provides vertical and horizontal liquid levels, vibrations, direction, and speed of wind in Quang Nam. All sensors are connected to their corresponding sinks; so sensed data is sent to the server through general packet radio service (GPRS) [8]. Nguyen-Son *et al.* have not done vulnerability detection in WSN used for two cities. At the same time, they have done good analysis of vendors, installation strategy, and performance analysis. Disaster area network (DAN) is very useful in the case of natural or man-made disasters [9]. Rapidly deployable networks are required for search and rescue operations during the state of extreme emergency due to disasters. The survey of [9] provides comparative study of network architecture and routing models for DANs. The main goal is to achieve a reliable, energy efficient, mobile network which can provide less delay, less overhead, energy efficiency, more bandwidth, and rapid mobility for multimedia applications. Mobile *ad hoc* networks (MANET) and integration of MANETs with other networks are proposed to meet the above requirements [9]. Sensor-based event detection is used to develop emergency alerting systems in different smart city zones [10]. Sensors detect particular events in the smart city zone and an event report is generated. Emergency alerts are generated in particular smart city zones based on the event report. The main advantage is the distributed nature of the emergency alert generation system in [10]. The prototyping was done by using Raspberry Pi interfaced with air quality sensor, temperature sensor, audio sensor, humidity sensor, water level sensor, GPS module, and display. The value of severity level is computed to indicate the emergency magnitude of the particular event. The vulnerability and anomaly can impact a system to generate false emergency alert in the smart city zones. The vulnerability and anomaly analysis of the emergency alert system proposed in [10] may help to deal with security threats. The violence detection using wireless video sensor network and deep learning is proposed for smart cities in [11]. Spatial and temporal information is processed in parallel to obtain real-time computationally efficient implementation using Raspberry Pi [11]. The approach of [11] achieves 86.93 with 0.21 standard deviation and area under curve (AUC) score of 0.9543. Anomaly and vulnerability analysis of sensor networks is important for the violence detection approach in [11]. A review on big data for visualizing, analyzing, and predicting natural disasters is proposed in [12]. Satellite imagery, unmanned aerial vehicle (UAV) based aerial imagery and videos, wireless sensor web and IoT, LiDAR, simulation data, vector-based spatial data, crowdsourcing, social media, mobile GPS, and call data records are the different data sources used for visualization, analysis, and prediction of natural disasters [12]. All disaster management phases are reviewed with supporting literature in [12]. The main challenges are data collection from diverse sources, management technologies, and developing efficient systems for disaster management with least losses in terms of human lives and economy [12]. Post-disaster IoT-based crowd sensing aware disaster management is proposed with crowd sensing clustering algorithm and fuzzy logic based decision support system [13]. Cluster-wise review and analysis are provided for different disaster management systems in [13].

Cellular and Wi-Fi based IoT architectures are proposed with detailed finite state machine analysis, building damage status, population density affected by disaster, and fuzzy logic based post-disaster decision system [13]. Overview and comparison of SENDROM (earthquake detection), INSIEME (emergency operations), telemedicine with WSN (victim evaluation), WINSOCK (landslide detection), USN4D (air pollution detection), AWARE (surveying and filming fire detection), and MiTag (tracking patients) are given with challenges in [14]. Earthquake alert generations based on animal behavior detection, ground water level variations, radon gas emission, increase in temperature around fault lines or earthquake prone areas using ICT infrastructure and WSN [15]. Anomaly and vulnerability in such earthquake alert systems may cause unnecessary chaos in the region; so analysis of such vulnerability is extremely important in this kind of system. The concept of smart city, its emergence, and role of integrated sensor systems for smart cities are discussed in [16]. Smart transport and mobility tracking, smart grid, smart environment, and sensors for information collection and information connection are discussed in great detail [16]. The vulnerability analysis can add extremely important aspects to smart city development. A detailed review of machine learning in IoT security is provided in [17]. Machine learning and deep learning approaches are reviewed for authentication, attack detection and mitigation, distributed DOS attack, anomaly/intrusion detection, and malware analysis [17]. Traditional network security concepts cannot be applied directly to IoT network due to resource constraint and distributed and diverse nature of it [17]. Machine learning and deep learning require large datasets for benchmarking security solutions, but it is difficult to find public dataset in this domain [17]. Next generation disaster data infrastructure is discussed including disaster data collection and transmission, disaster data processing, disaster data quality control, and visualization in [18]. Predictive data analytics, real-time location aware information, and natural disaster patterns are extremely important to respond to the natural disaster crisis [18]. The future of smart cities is predicted for autonomous vehicles, positive train controls, intelligent transportation systems, vehicle-to-vehicle and vehicle-to-infrastructure, smart power generation plants, smart distribution and transmission, advanced metering infrastructure, smart water treatment, smart water distribution, and smart water storage with cyber–physical infrastructure security risk and opportunities in [19]. Extensive survey is done on data management, security and enabling technologies in smart cities with smart street light, smart health, smart traffic management, smart emergency system, virtual power plants, data acquisition technologies knowledge discovery, data presentation, data dissemination methods and applications, smart city requirements and evaluation, cloud computing, network functions virtualization, and software defined networking [20]. Sensor networks, unmanned aerial vehicles, mobile *ad hoc* networks, social networks, vehicular *ad hoc* networks, crowdsourcing, device-to-device communication, 5G, IoT, machine learning, deep learning, real-time analytics, direct access, opportunistic routing, different algorithms, and platforms are reviewed as important technologies for actual development of smart city [20]. Smart systems are proposed for disaster prediction, discovery and response in smart cities with main building blocks of smart cities, generic people-centric, IoT-centric sensing architectures and main challenges [21]. WSNs in depth review is described with introduction, history, and industrial drivers of WSNs, WSN technology, challenges of WSNs, WSN applications in the infrastructure systems, and standards of WSNs and systems in [22]. Smart cities and emergency services are proposed with smart environment, smart mobility, smart living, smart people, smart economy, smart governance, video surveillance and public security, pollution control, flood alert system, and smart city platform four layers in [23]. A detailed research on smart fighter's fighting is done with stationary and mobile sensors, data collection, hardware, software, real-time analytics, and pre- and post-emergency scenario applications in [24]. Different algorithms are compared for anomaly detection in smart city WSNs [25]. Real data is taken from Barcelona for validation of concept and one-class support vector machine outperforms other techniques in anomaly detection [25]. Cyber–physical vulnerabilities like routing attacks (sybil wormhole, sinkhole, selective forwarding, etc.), service denial attacks (denial of service, distributed denial of service, flooding, etc.), insider attack, cyber intrusion, physical destruction, physical tampering, environment tampering, and physical intrusion are reviewed with different solution approaches in smart city WSNs [26]. One class quarter sphere support vector machine algorithm is proposed to detect anomalous measurements in the sensor data from great duck island project [27].

WSN protocol stack, security requirements, and different attacks at each layer of protocol stack are reviewed in [28]. WSN security issues like key management, secure time synchronization, secure location discovery, etc. still have more scope for research activities [28]. Sleep deprivation attack (which can drain battery of WSN nodes and reduce battery life) detection is proposed with a two-step approach to reduce probability of the false intrusion detection in [29]. A very good survey is written on potential security risks of smart cities including different vulnerabilities, risks, mitigation and prevention [30]. Different vulnerability and security attacks are classified according to the presence and intervention power dimensions based on generic adversary models in [31].

9.2 Proposed Analysis

The proposed sensor network consists of mobile as well as fixed nodes representing entities capable of tracking infection spreading, event propagation, different radio technologies, geographical and topological considerations, and battery performance. CupCarbon [32] is an open source tool for simulating smart city and IoT WSNs. It facilitates the design and visualization of sensor networks with inclusion of various environmental scenarios such as fires, gas, mobiles, etc. It also helps researchers to test different wireless network topologies and protocols. Given a fixed size geographical area, nodes are placed to form a WSN. For the present study, a test area of approximately 4 sq. km is considered. All fixed nodes are capable of handling multiple radio technology transmissions such as LoRa, Wi-Fi, etc. Mobile nodes interact with fixed nodes over ZigBee as soon as they come in communication range of fixed nodes. The information regarding node interaction is sent to a central repository (CR) over LoRa. Figure 9.1 shows the network shape with placement of fixed nodes. The number of nodes and their placement can be varied in a network in order to study the performance of the network.

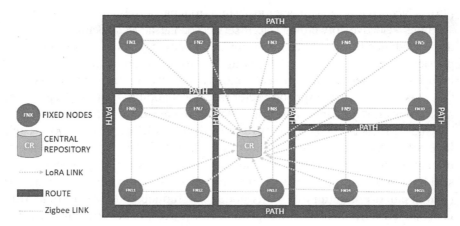

Figure 9.1   Test network setup.

9.3 Modeling and Simulation

In this study, it is assumed that certain mobile nodes would be tagged as "infected" depicting a person carrying infection. Under real-life situation, this can be compared with performing localization operation. Simulations are performed for proposed test network considering the following scenarios:

(i) Movement of multiple mobile nodes across network under normal conditions.

(ii) Movement of "infectious" nodes.

(iii) Node parameter comparison and analysis for these cases.

(iv) Analysis under condition of random path selection.

Figure 9.2 shows the CupCarbon simulation environment with test network consisting of 15 nodes. The arrow links represent the state of communication between nodes. All these nodes are configured as fixed nodes. Fixed nodes are configured in "receive" mode. As soon as they sense any movement from mobile node, the information is transmitted to CR. The mobile node is not shown in this figure as it is assumed that it will be identified with its motion throughout the network.

The flow chart for the network flow is shown in Figure 9.3. While fixed nodes are provided with ZigBee support for communicating with mobile nodes and inter-communication between fixed nodes, LoRa support is provided for direct communication with CR. This makes sure that every event is captured

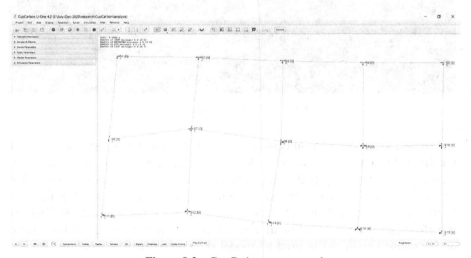

Figure 9.2 CupCarbon test network.

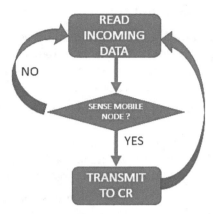

Figure 9.3   Fixed node task.

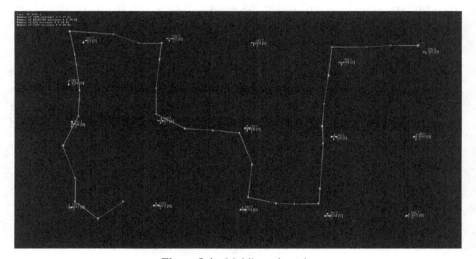

Figure 9.4   Mobile node path.

and subsequently recorded in central server. Figure 9.4 shows the route selected for the movement of mobile node. As soon as the mobile node starts to move, it broadcasts a unique signal which is received by the fixed nodes. Fixed nodes are programmed in such a manner that a unicast message is sent to CR through LoRa. In this way, track of mobile node movement is kept. Again, the simulation is performed for various hyper-parameters and the same is investigated for battery power consumption. AI is used to model different scenarios based on battery power consumption.

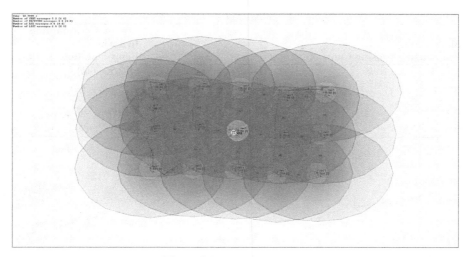

Figure 9.5 Node interaction.

Figure 9.5 depicts visual representation of the interaction between mobile node and node S8. In real-life scenario, multiple routes can be configured and the mobile node can be made to randomly choose a particular path. The choice made by mobile node affects the network performance.

9.4 Results and Discussions

Table 9.1 comprises details of hyper-parameters for various simulation campaigns, and their impacts on energy consumption are shown in Figure 9.6. These parameters include device parameters such as sensing constant, UART data rate and drift (sigma), and simulation speed.

The potential choices for application of AI for such a WSN can be inclusive of both regression and classification. For example, for given device parameters, the rate of depletion of energy for individual nodes can be predicted and any deviation can give us an idea about the state of the network functioning. The most widely used AI techniques include Naive Bayes, decision trees, random forests, and artificial neural networks (ANNs).

9.4.1 Application in Pandemic Control

Since the idea is to apply the proposed solution for pandemic control, it would be easier to devise a strategy where the node ID of a particular

Table 9.1   Simulation parameter values

Simulation Campaign	Sensing Constant	Drift (sigma)	Energy Consumption Behavior Representation
1	1	3×10^{-5}	Figure 9.6(a)
2	1	4×10^{-5}	Figure 9.6(b)
3	2	3×10^{-5}	Figure 9.6(c)
4	2	4×10^{-5}	Figure 9.6(d)

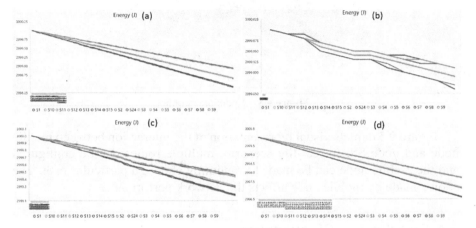

Figure 9.6   Energy consumption pattern.

mobile node would be mapped to a person who is supposed to always carry the hand-held node device. This will not only ensure smooth tracking but also the network state would give us clues about any deviation in parameters. ANN can be used to produce intelligent networks, where nodes can re-configure their device and radio parameters, thereby improving performance. This could also help tackle situations where there are densely populated containment zones.

9.5 Conclusion

A WSN was proposed for tackling global emergency situations such as occurrence of pandemics. The idea behind envisaging a strategy of WSN deployment for pandemic control is to efficiently monitor and control the spread of disease. A CupCarbon-based simulation of network consisting of 15 nodes including fixed and mobile nodes was discussed. The biggest advantage of using such tools is that it helps understand network behavior without building them in real time and thereby saving cost. The future work for such analysis could include in-depth analysis of AI techniques for their performance.

References

[1] Eleana Asimakopoulou and Nik Bessis. Buildings and crowds: Forming smart cities for more effective disaster management In Innovative Mobile and Internet Services in Ubiquitous Computing (IMIS), 2011 Fifth International Conference on, pages 229–234. IEEE, 2011.

[2] SyedIjlal Ali Shah, Marwan Fayed, Muhammad Dhodhi, and Hussein T Mouftah. Aqua-net: a flexible architectural framework for water management based on wireless sensor networks. In Electrical and Computer Engineering (CCECE), 2011 24th Canadian Conference on, pages 000481–000484. IEEE, 2011.

[3] Hsu-Yang Kung, Jing-Shiuan Hua, and Chaur-Tzuhn Chen. Drought forecast model and framework using wireless sensor networks. Journal of information science and engineering, 22(4):751–769, 2006.

[4] Alberto Rosi, Matteo Berti, Nicola Bicocchi, Gabriella Castelli, Alessandro Corsini, Marco Mamei, and Franco Zambonelli. Landslide monitoring with sensor networks: experiences and lessons learnt from a real-world deployment. International Journal of Sensor Networks, 10(3):111–122, 2011.

[5] Maneesha Vinodini Ramesh. Design, development, and deployment of a wireless sensor net- work for detection of landslides. Ad Hoc Networks, 13:2–18, 2014.

[6] Xia M., Song D. (2018) Application of Wireless Sensor Network in Smart Buildings. In: Gu X., Liu G., Li B. (eds) Machine Learning and Intelligent Communications. MLICOM 2017. Lecture Notes of the Institute for Computer Sciences, Social Informatics and Telecommunications Engineering, vol 226. Springer, Cham.

[7] Minoli, D., Sohraby, K., Occhiogrosso, B.: IoT considerations, requirements, and architectures for smart buildings-energy optimization and next-generation building management systems. IEEE Internet Things J. 4, 269–283 (2017). IEEE Press.

[8] Vo, Nguyen-Son, Antonino Masaracchia, Long Dinh Nguyen, and Ba-Cuong Huynh. "Natural Disaster and Environmental Monitoring

System for Smart Cities: Design and Installation Insights." EAI Endorsed Trans. Indust. Netw. & Intellig. Syst. 5, no. 16 (2018): e5.

[9] Jahir, Y., Atiquzzaman, M., Refai, H., Paranjothi, A. and LoPresti, P.G., Routing protocols and architecture for disaster area networks: A survey. Ad Hoc Networks, 82, pp.1-14,2019.

[10] Costa, D.G., Vasques, F., Portugal, P. and Aguiar, A., 2020. A Distributed Multi-Tier Emergency Alerting System Exploiting Sensors-Based Event Detection to Support Smart City Applications. Sensors, 20(1), p.170.

[11] Baba, M., Gui, V., Cernazanu, C. and Pescaru, D., 2019. A sensor network approach for violence detection in smart cities using deep learning. Sensors, 19(7), p.1676.

[12] Yu, M., Yang, C. and Li, Y., 2018. Big data in natural disaster management: a review. Geosciences, 8(5), p.165.

[13] Kucuk, K., Bayilmis, C., Sonmez, A.F. and Kacar, S., 2019. Crowd sensing aware disaster framework design with IoT technologies. Journal of Ambient Intelligence and Humanized Computing, pp.1-17.

[14] Benkhelifa, I., Nouali-Taboudjemat, N. and Moussaoui, S., 2014, May. Disaster management projects using wireless sensor networks: An overview. In 2014 28th International Conference on Advanced Information Networking and Applications Workshops (pp. 605-610). IEEE.

[15] Ur Rahman, M., Rahman, S., Mansoor, S., Deep, V. and Aashkaar, M., 2016. Implementation of ICT and wireless sensor networks for earthquake alert and disaster management in earthquake prone areas. Procedia Computer Science, 85, pp.92-99.

[16] Ba i , Ž., Jogun, T. and Maji , I., 2018. Integrated sensor systems for smart cities. Tehni ki vjesnik, 25(1), pp.277-284.

[17] Hussain, F., Hussain, R., Hassan, S.A. and Hossain, E., 2020. Machine learning in IoT security: current solutions and future challenges. IEEE Communications Surveys & Tutorials.

[18] LODGD (2019), Next Generation Disaster Data Infrastructure -White Paper. CODATA Task Group, Linked Open Data for Global Disaster Risk Research (LODGD). September 2019. Paris.

[19] RISK, C.P.I., 2015. THE FUTURE OF SMART CITIES: CYBER-PHYSICAL INFRASTRUCTURE RISK.

[20] Gharaibeh, A., Salahuddin, M.A., Hussini, S.J., Khreishah, A., Khalil, I., Guizani, M. and Al-Fuqaha, A., 2017. Smart cities: A survey on data management, security, and enabling technologies. IEEE Communications Surveys & Tutorials, 19(4), pp.2456-2501.

[21] Boukerche, A. and Coutinho, R.W., 2018, June. Smart disaster detection and response system for smart cities. In 2018 IEEE Symposium on Computers and Communications (ISCC) (pp. 01102-01107). IEEE.

[22] Yinbiao et al., "Internet of Things: Wireless Sensor Networks", IEC White Paper, September 12, 2019. https://www.iec.ch/whitepaper/internetofthings/.

[23] Astilleros et al., "Smart cities and Emergency Services", EENA Operations Document, September 2016.

[24] Grant et al., "Research Roadmap for Smart Fire Fighting", National Institute of Standards and Technology Special Publication 1191, 246 pages, May 2015. www.nfpa.org/SmartFireFighting.

[25] Garcia-Font, V., Garrigues, C. and Rifà-Pous, H., "A comparative study of anomaly detection techniques for smart city wireless sensor networks", sensors, 16(6), p.868, 2016.

[26] Hasan, M.M. and Mouftah, H.T., "Cyber-physical vulnerabilities of wireless sensor networks in smart cities", Security and Privacy in Cyber-Physical Systems: Foundations, Principles, and Applications, 2017.

[27] Rajasegarar, S., Leckie, C., Palaniswami, M. and Bezdek, J.C., "Quarter sphere based distributed anomaly detection in wireless sensor networks" In 2007 IEEE International Conference on Communications (pp. 3864-3869). IEEE, June 2007.

[28] Kavitha, T., and D. Sridharan. "Security vulnerabilities in wireless sensor networks: A survey." Journal of information Assurance and Security 5, no. 1, 31-44, 2010.

[29] Bhattasali, Tapalina, Rituparna Chaki, and Sugata Sanyal. "Sleep deprivation attack detection in wireless sensor network." arXiv preprint arXiv:1203.0231 (2012).

[30] Kitchin, Rob, and Martin Dodge. "The (in) security of smart cities: Vulnerabilities, risks, mitigation, and prevention." Journal of Urban Technology 26, no. 2, 47-65, 2019.

[31] Benenson, Zinaida, Peter M. Cholewinski, and Felix C. Freiling. "Vulnerabilities and attacks in wireless sensor networks." Wireless Sensors Networks Security, 22-43, 2008.

[32] CupCarbon. A Smart City & IoT Wireless Sensor Network Simulator. http://labsticc.univ-brest.fr/~bounceur/cupcarbon/doc/cupcarbon_user_guide.pdf (Accessed online : 25-08-2020)

10

Peculiarities of Technical Measures During the COVID-19 Pandemic

Iosif Z. Aronov[1], Anna M. Rybakova[1], Nataliia M. Galkina[2]

[1]Moscow State Institute of International Relations University, Moscow, Russia
[2]International Trade and Integration (ITI) Research Center, Moscow, Russia
Corresponding Author: Nataliia M. Galkina, nmgalkina@itandi.ru

Abstract

The COVID-19 pandemic has become the most significant global crisis in recent history, forcing World Trade Organization (WTO) member states to focus on protecting lives, public health, and food security.

In the face of the pandemic, WTO member states used a variety of government regulation and trade protection measures, but the role of technical barriers to trade (TBT) measures proved to be significant. According to the WTO data, about 38% out of 198 trade measures notified to the WTO from 1 February 2020 to 1 July 2020 were TBT measures (sanitary and phytosanitary (SPS) measure notifications amounted to 28%).

This chapter provides an overview of technical measures undertaken by WTO member states during this pandemic.

10.1 Introduction

The COVID-19 pandemic has become the most significant global crisis in recent history that required states to focus on protecting human lives, ensuring public health and food security of the population. World Trade Organization (WTO) members took trade-related measures in response to the pandemic; however, the terms "coronavirus," "COVID," "SARS-CoV-2," and "nCoV" have begun to be used in notification texts concerning the adoption of non-tariff trade regulatory measures, identifying the purposes and reasons for the adoption thereof.

According to the WTO, as of 1 July 2020, more than 70 countries with varying activities applied non-tariff regulatory tools to combat the pandemic (Figure 10.1).

The maximum number of measures related to trade regulation during the pandemic was taken by Brazil – 25 measures; Kuwait – 16 measures, Philippines and Thailand – 11 measures; Colombia, Republic of Korea, and EU – 8 measures; Indonesia, Israel, and Ukraine – 7 measures each; Argentina, Australia, Egypt, Paraguay, Peru, and USA – 6 measures each. The Russian Federation, Canada, the Dominican Republic, and Switzerland reported on the five COVID-19-related measures taken.

In the face of the pandemic, the WTO members applied various non-tariff measures of government regulation and trade protection (Figure 10.2).

Among these measures are Council for trade in goods, export restrictions, market access for goods, quantitative restrictions, including import/export quotas, sanitary and phytosanitary measures, technical measures or technical barriers to trade, trade facilitation agreement, trade-related aspects of

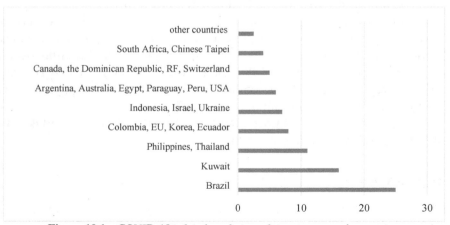

Figure 10.1   COVID-19-related market regulatory measures by country.

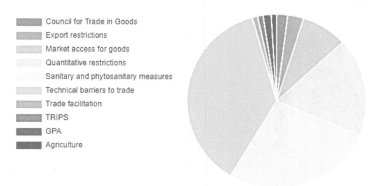

Council for Trade in Goods
Export restrictions
Market access for goods
Quantitative restrictions
Sanitary and phytosanitary measures
Technical barriers to trade
Trade facilitation
TRIPS
GPA
Agriculture

Figure 10.2 Distribution of notifications concerning measures taken in response to the pandemic by measure type (Source: WTO).

intellectual property rights, Agreement on Government Procurement, and measures to support (restrict) agricultural trade.

It is obvious that the largest number of notifications was devoted to the application of SPS measures and TBT measures. The first pandemic-related notification appeared on 2 February 2020 and was posted by the Russian Federation that banned the import of certain animals and insects from the PRC.

It is worth noting that among 190 trade measures, notifications of which were posted in the WTO bases from 1 February 2020 to 1 July 2020, notifications on TBT measures accounted for 40% (notifications on SPS measures accounted for 30%) [1].

10.2 Application of TBT Measures By WTO Members

In this article, the results of the application of TBT measures are presented in the face of the COVID-19 pandemic for the period from 1 February 2020 to 1 July 2020, including monitoring of notifications submitted by WTO members in ePing – SPS and TBT Notification Alert System, as well as monitoring of actions taken by international, regional, and national standardization authorities to prevent the spread of infection and cope with the impacts of the pandemic.

The WTO TBT Agreement [2] provides for the possibility of introducing emergency TBT measures. As has been shown in practice, the application of such TBT measures has proved to be a necessary tool for market regulation in many countries during the pandemic.

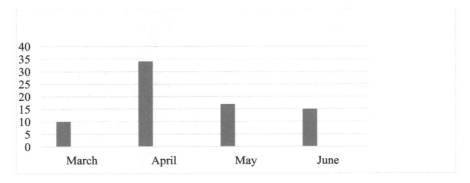

Figure 10.3   Distribution of notifications concerning COVID-19-related TBT measures by month.

The first notification of the TBT measure taken to combat the coronavirus was posted by Brazil on 16 March 2020 and provided information on the approval by the INMETRO[1] of Ordinance No. 79 dated 4 March 2020 [3] on the Introduction of Extraordinary Conditions for Execution of Conformity Assessment Activities in Countries Affected by Coronavirus Epidemic, including by means of "contactless" assessment.

In March–June 2020, the total number of posted notifications concerning TBT measures was 76 [4]; however, the peak of notifications was attained in April 2020 (Figure 10.3), when 34 notifications were posted.

A content analysis of notifications revealed that TBT measures were taken in relation to certain groups of critical goods in a crisis, namely medical devices, means of protecting populations (respirators, masks, and sanitizers), diagnostic test kits to detect the virus, medicines, including active substances, serum, vaccines, as well as food products and materials for their packaging.

Further, Table 10.1 shows the results of the notification analysis.

Brazil posted the maximum number of notifications – 21 notifications. The TBT measures taken were related to the simplification of the requirements for the production of medicines and pharmaceutical substances, facilitation of registration and obtaining the import permits for medical devices, as well as certification facilitation.

The notifications posted by Kuwait provide information on the adoption of 16 mandatory standards containing requirements for medical devices, including equipment, masks, sanitizers, and methods for the assessment

---

[1] The National Institute of Metrology, Standardization and Industrial Quality.

Table 10.1 Notifications concerning TBT measures posted during March–June 2020.

WTO member	Number of notifications concerning TBT measures	Content of notifications
Brazil	21	Facilitation of registration, verification, obtaining approvals for the import of medical devices, medicines, protective equipment; suspension of certification relating to surgical gloves, certification of products and services in "contactless" mode
Kuwait	16	Adoption of standards for medical equipment, masks, etc., for the test methods thereof
Thailand	7	Certification of each batch; requirements for working with hazardous substances, facilitation of registration of medical devices
Ecuador	3	Application of electronic certificates of free sale for import, requirements for disinfectants
Indonesia	3	Suspension of certain food standards requirements
USA	3	Establishment of requirements and test methods for respirators, clarification of laboratory designation procedures
Uganda	3	Adoption of mask standards
Morocco	2	Introduction of mandatory requirements for masks and protective screens, certification of these products
Argentina	2	Introduction of mandatory requirements for protective equipment and the certification thereof
Republic of Korea	2	Requirements for masks, the use of medical masks in daily life, restrictions on polymer waste import
Taiwan	2	Requirement for inspection and examination of imported masks
Switzerland	2	Facilitation of medication and food import
UAE	1	Change in the certification procedure relating to imported goods (batch certification)
Namibia	1	Adoption of the disinfectant standard
Czech Republic	1	Determination of the maximum purchase prices for medical devices
Canada	1	Temporary facilitated access to the mask, disinfectant market
Peru	1	Establishment of requirements for masks
Kenya	1	Establishment of requirements for used clothing
Ukraine	1	Facilitation of the conformity assessment of medical devices, masks, respirators
Jamaica	1	Adoption of the disinfectant standard
Group of countries[1]	3	Request for suspension of the EU requirements (plant protection products in food of plant origin)

---

[1] Argentina, Benin, Burkina Faso, Cape Verde, Colombia, Costa Rica, and others.

and testing thereof. For the first time, the transition period was reduced in the notifications concerning the adoption of standards (Kuwait) on an emergency basis – standards shall become effective 7 days after their official publication [5].

Uganda, Morocco, Argentina, Namibia, Peru, Jamaica, USA, Ecuador, and Republic of Korea also gave notification for critical goods. Depending on the product requirements in these countries, the notifications related to the adoption of mandatory standards or regulatory documents that include such requirements directly or by referring to standards.

Notifications *concerning the introduction of product requirements* were mainly related to the means of protecting populations such as respirators, masks and protective screens, disinfecting cosmetics, etc. Measures associated with mandatory requirements were introduced by all countries on an ongoing basis and in record time. An analysis of notifications concerning TBT measures enables to make conclusions that insufficient attention has been paid by most developing countries to the means of protecting populations until 2020.

At the same time, measures have been introduced *suspending mandatory requirements* for a specific period or until further notification. For example, Indonesia, in relation to food products, notified the WTO members of a temporary refusal to apply mandatory requirements for food products [6]. Switzerland has adopted the Ordinance on Food and Common Products, which temporarily reduces the requirements for food ingredients and packaging materials that may be in short supply during the pandemic. Such food products must be marked with a special sign (a red round sticker) [7].

Many countries have adopted measures related to *the acceleration and facilitation of conformity assessment* and market access for critical goods. These measures included the use of electronic certificates of free sale (Ecuador) [8]; electronic registration services (Thailand, etc.) [9]; reduction of requirements and list of documents for registration (Canada, Brazil, etc.) [10, 11]; introduction of an emergency procedure for importing a number of medical devices, the compliance of which with the mandatory requirements has not been confirmed (Ukraine) [12]; temporary refusal to carry out compulsory certification (Brazil) [13].

The need to apply quarantine measures during the pandemic has led to *certification facilitation during the pandemic* in several countries. While Brazil allows the issuance of certificates of conformity without verification and testing by a third party, based on the acceptance of the manufacturer's test results [14], Thailand [15] and the UAE temporarily apply only one certification scheme for imported goods such as certification of each batch of incoming consignments.

The establishment of international arrangements is another feature of TBT measure during the pandemic. Singapore–New Zealand Declaration on Trade in Essential Goods for Combating the COVID-19 Pandemic [16], published on 16 April, proposed not only to abolish customs duties and lift export bans but also to reduce TBT restrictions on critical and essential goods to combat the pandemic. The declarations of the ASEAN [17] and APEC [18] countries also contain a refusal to introduce TBT measures and their reduction, including through the use of digital tools and platforms. The notifications submitted by the group of countries, i.e., requests to suspend the EU requirements for the content of plant protection products in food of plant origin in order to maintain food supply chains, fit well therewith [19].

Thus, following the analysis of notifications concerning TBT measures taken in response to the COVID-19 pandemic, we may therefore note several trends:

- urgent adoption of the necessary mandatory requirements (regulatory documents or standards) for critical goods during this period in countries where these requirements do not exist;
- temporary refusal to apply mandatory requirements (for example, for food products) and certification;
- temporary facilitation of conformity assessment of medical devices and protective equipment (reducing the time needed to conduct inspections, using electronic copies of documents required for the purpose of completing formalities);
- the application of "contactless" certification procedures, including the use of the manufacturers' test results and reports to issue certificates of conformity;
- the creation of opportunities for electronic document filing to register goods and obtain import permits.

Not all the COVID-19 pandemic-related TBT measures taken by States were reflected in the notifications sent to the WTO bases. Thus, the Russian Government, with its Decree No. 430 dated 3 April 2020, introduced measures aimed at facilitating the registration of medical devices for emergency use (ventilators, test systems and antibody test kits, protective clothing, masks, respirators, gloves, and thermometers), including electronic document filing, the use of foreign test protocols, and the sale of disposable products unregistered in Russia; however, this was not reported [20]. Obviously, such practice is followed not only in Russia.

In March 2020, Commission Recommendation (EU) 2020/403 dated 13 March 2020 on Conformity Assessment and Market Surveillance Procedures within the Context of the COVID-19 Threat [21] were issued which recommend EU Members to examine the possibility of deviation from the conformity assessment procedures for medical devices and personal protective equipment. Perhaps, Ukraine[2] followed these recommendations, as it allows the import of medical devices, diagnostic agents and protective equipment that have not passed the conformity assessment in accordance with the requirements of European directives from 1 April 2020.

It should be noted that the simplification of registration and import requirements for medical devices and means of protecting populations may have an expected negative effect; the reduction of requirements may weaken the security, as indirectly indicated, in particular, by the decision made by Taiwan in June 2020 to conduct examination and inspection of imported masks (notification G/TBT/N/TPKM/422 dated 30 June 2020).

10.3 Peculiarities of Application of Standardization Tools During the Pandemic

The analysis of notifications concerning TBT measures taken in response to the COVID-19 pandemic shows that, roughly, half of the measures were related to standards (overall 35 notifications referring to standards, Figure 10.4).

Most of the adopted standards establish requirements for masks, hygiene products (sanitizers), ventilators, and other medical devices. It is clear that emergency standards are needed to ensure the safety, reliability, and durability of critical goods such as medical products and means of protecting populations; the application of international or foreign standards as a basis for national standards stimulates the barrier-free exchange of goods.

We can agree with the fact that apart from standards containing requirements for critical goods and methods for the testing thereof, the role of other standards will continue to increase in an emergency situation [22]:

- those containing requirements for quality management of manufacturers, for example, ISO 13485 "Medical devices. Quality management systems. System requirements for regulatory purposes";

---

[2] The EU and Ukraine signed the Association Agreement in 2014.

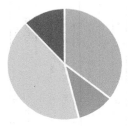

- adoption of standards
- approval of requirements based on standards
- facilitation of conformity assessment
- other measures

Figure 10.4   TBT measures related to various technical regulatory tools.

- those ensuring the quality of work of medical laboratories (such as ISO 15189 "Medical laboratories. Requirements for quality and competence");
- those containing recommendations and guidelines for ensuring business continuity and stability (such as ISO 22301 "Business continuity management systems. General requirements").

10.3.1 Standard Accessibility in Response to the COVID-19 Pandemic

The texts of international, regional, and national standards are usually provided by standardization authorities for a fee; the funds received support the activities of standardization authorities. This approach in the face of the pandemic has become a restrictive condition for the use of standards, in particular, by enterprises rebuilding production to protect against the pandemic.

The first tool applied by standardization authorities was the provision of free access to the texts of the standards (see Table 10.2).

It is characteristic that the first national standardization authority to provide such access to the necessary standards on 18 March was UNI – the National Standards Service of Italy, where, in March 2020, the epidemic became threatening. Then, in response to a request from the European Commission, on 20 March 2020, CEN, the European Committee for Standardization,

Table 10.2 Ensuring access to standards in a pandemic.

Standardization authority	Standards provided for free access	Comments
International Organization for Standardization (ISO)	22	Standards for equipment, medical masks, personal protective equipment, emergency management, and business continuity
International Electrotechnical Commission (IEC)	5	Standards for medical electrical devices
European Committee for Standardization and European Committee for Standardization in Electrical Equipment (CEN, CENELEC)	14	Standards for masks, protective clothing, and medical electrical devices
Italian National Standards Service (UNI)	27	Free access was granted on 18 March
British Standards Institute (BSI)	119	The standards are given in sections, and links to standards of other organizations for standardization are added
German Standards Institute (DIN)	38	Special publications of standards have been prepared
Netherlands Institute of Standards (NEN)	38	The standards are given in sections
French Standards Association (AFNOR)	25	National and international standards for medical devices, protective clothing, and medical masks
Swedish Institute of Standards (SIS)	44	The standards are available for reading; downloading for a reduced fee; communication with the TC for standardization that developed the standards is provided
Latvian authority for standardization (LVS)	62	National standards that are harmonized with international or European standards, as well as texts of international standards
Saudi Organization for Standards, Metrology and Quality (SASO)	32	The standards are given in sections
Singapore Standards Council (SSC)	40	The standards are given in sections (Singapore standards, ISO standards, standards for masks, etc.); there are links to standards of other organizations
Association of Technical Standards of Brazil (ABNT)	32	Standards for protective equipment, materials for their manufacture, and medical equipment
Agency for technical regulation and metrology of the Russian Federation (Rosstandart)	52	National and international standards for medical devices, protective clothing, and medical masks, risk assessment standards
Standardization Administration of China (SAC)	Not revealed	Lists of national standards and the results of an analysis of the equivalence of standards in English have been published; English texts can be obtained from the main catalog, for a fee
Indian Bureau of standards (BIS)	Three standards for protective equipment	All mandatory standards are publicly available for viewing Guides on mask application and workstations are published

and CENELEC, the European Committee for Standardization in Electrical Engineering, decided to place a series of European Standards (EN) for medical devices and personal protective equipment used in the context of the COVID-19 pandemic in the public domain. The international organizations for International Organization for Standardization (ISO) and International Electrotechnical Commission (IEC) standardization made a similar decision in early April, and on 16 April 2020 on the ISO website in the section dedicated to the response to the COVID-19 pandemic, 22 standards have been published (for medical devices and their control methods, risk assessment, emergency management, and business sustainability standards), the texts of which are available by reference to the standard designation. Since then, most national standards authorities have taken similar actions; Table 10.2 provides data on 13 analyzed national standardization authorities [23, 24].

The "open" standards include national standards that are harmonized with ISO and IEC standards and national standards that are harmonized with European standards, as well as a number of national standards for the same standardization objects and texts of international standards themselves.

As of 1 July 2020, the maximum number of standards in the public domain is available on the website of the British Standards Institute (BSI) – 119 standards are in searchable thematic sections (Table 10.2), including standards for testing methods of virulent activity of disinfectants and standards for materials for the manufacture of protective clothing. BSI, by agreement with other organizations, provides access to their standards and publications (AAMI[3], IEEE[4], and ASTM[5]) [25].

It is interesting that the website of the Standardization Administration of China (SAC) published the results of an analysis of the equivalence of Chinese and EU standards for masks (GB 0469-2011 "Surgical masks" and EN 14683-2019 "Surgical masks – requirements and test methods") [26]. Although not all values of indicators are identical, the conclusions about the equivalence of the PRC and CEN (EN) standards led to the approval of the supply of masks produced in China by the regulatory authorities of a number of European countries (the Netherlands, France, Germany, and Spain).

With regard to the conditions for accessing standards in this study, there are three ways to get acquainted with the texts of standards:

---

[3]The Association for the Development of Medical Instruments.
[4]The Institute of Electrical and Electronics.
[5]The American Society for Testing and Materials.

1. direct access to the text via a link on the website of the standardization authority (ISO and ABNT);
2. access to the text after notification registration (BSI and DIN);
3. access after sending a request (Rosstandart and LVS).

Despite some differences, all these methods allow getting acquainted with the texts of standards quickly enough, without additional time costs.

10.3.2 Development of Standardization Documents in Response to the COVID-19 Pandemic

There are two trends in the development of standardization documents in response to the COVID-19 pandemic:

1. development of national standards as a matter of urgency;
2. development of incomplete consensus documents.

Examples of the first trend are the urgent development of national standards by the national standardization authorities of Kuwait, Jamaica, Uganda, and Spain (Table 10.3).

The mandatory standards adopted in Kuwait, Jamaica, and Uganda were not only developed as a matter of urgency but also came into effect without the usual two-month transition period, which means that national standardization authorities have the appropriate authority to deal with emergencies.

The second way to urgently develop requirements and recommendations during a pandemic is the adoption of incomplete consensus documents by standardization authorities. Many European bodies, including CEN, have followed this way.

Incomplete consensus documents, in contrast to standards, can be developed in a much shorter time since their development does not require a procedure for reaching consensus (general agreement), and even public discussion is not always necessary. The main types of incomplete consensus

Table 10.3 Examples of accelerated adoption of national standards during the COVID-19 pandemic.

Country/standardization authority	Standards	Comments
Kuwait/public service for industrial standards and metrology of Kuwait	Seven mandatory standards for masks, medical devices, and test methods	Effective 7 days after approval and publication
Uganda/National Bureau of Standards of Uganda	Three mandatory standards for respirators and masks	Effective immediately upon approval
Jamaica/National Bureau of Standards of Jamaica	One mandatory standard for hand sanitizers	Effective immediately upon approval

Table 10.4   Types of incomplete consensus documents developed during the pandemic.

Document	Adoption purpose	Adoption features	Development time
Guidance (BS, EN, ISO, IEC, etc.)	Provides general guidance with recommendations and reference information	Development and adoption by the technical committee after public discussion	12–15 months
Technical requirements (SPEC specifications)	A standardization document developed if there is no sufficient basis for a consensus standard. Shall not contradict the existing standards	Development and adoption by the technical committee or group of experts after public discussion	3 months
International workshop agreements (CWA/IWA)	Agreements of experts with the support of the industry or the States. CWA is an international (ISO, CEN) agreement.	Development and approval by the expert group	12 months

documents that were adopted by standardization authorities during the pandemic are shown in Table 10.4.

It is shown that the first object for the development of such documents were masks used by the population in the mode of self-isolation, the requirements for which are not regulated by the EU's EU's harmonizing legislation and national standards. Given that hygiene masks for a wide range of consumers have become a fundamental element of national strategies to overcome the pandemic, there was a need for their mass production, and the urgent development of documents that standardize the requirements for their safety and reliability was a timely response to COVID-19. On 27 March 27, 2020, the first document on standardization – AFNOR Spec-Protective (Barrier) Masks [27] was published, free access to which was provided by the national standardization authority of France.

This document is based on criteria approved by 150 experts and can be used by manufacturers. It was initially stressed that the document has a temporary status and will be reviewed. Similar documents have been developed by UNE (the Spanish Association of Standards) and the National Standards Authority of the Netherlands (Table 10.5). It is characteristic that until now, Netherlands Institute of Standards (NEN) has not used specification development as a tool for national standardization; NEN specifications were developed for the first time [28]. The European Association for Standardization CEN in June 2020 also published an incomplete consensus document CWA 17553:2020 *Face Masks. A Guide to Minimum Requirements, Test Methods, and Use* [29], containing minimum requirements for the design, production, and performance evaluation of face masks (barrier masks) intended for consumers, whether disposable or reusable.

Table 10.5 Development of incomplete consensus documents during the COVID-19 pandemic.

Country/standardization authority	Documents	Comments
France/AFNOR	AFNOR Spec	Requirements and templates for manufacturing protective (barrier) masks
Spain/UNE	Specifications: 0064-1 0064-2 0065 0066 series (21 documents)	Specifications 0064-1, 0064-2, and 0065 contain requirements for the manufacture, verification, and use of masks for adults and children, disposable and reusable; The 0066 series includes 21 guidelines for the implementation of activities in the field of tourism
Netherlands/NEN	Specifications: NEN spec NEN spec 2 NEN spec 3	Specifications are devoted to requirements for non-medical masks, ergonomic requirements for working premises during a pandemic, guidelines for the reuse of disposable medical devices
United Kingdom/BSI	Guidelines for the safe work during a pandemic COVID-19	The guidelines contain recommendations for workplace safety; version 2 of the document was published in June
EC/CEN	CWA 17553: 2020	Contains guidance on minimum requirements, test methods, and the use of non-medical face masks

It should be noted that national standardization authorities have also used the urgent development of incomplete consensus documents in order to establish requirements for safe work in the new environment (BSI and NEN) and for activities in a certain area of the economy (UNE).

The second version of the *BSI Guidelines for Safe Work During the Pandemic COVID-19* [30], updated and published in June, contains practical recommendations for organizing work, including safe placement of employees, communication rules, remote access, management rules in case of detection of diseases and rules for the safe use of sanitary zones, as well as providing psychological support for employees. Given that the document is available for download on the BSI website, it can be used in any organization and country in the world.

It is important to note another area of activity of national standardization authorities – the publication on websites of answers to questions related to coronavirus, and special reports containing reviews of practices.

Summing up the results of the study of the activities of standardization authorities in response to the COVID-19 pandemic, it can be stated that standardization authorities actively participated in the implementation of national emergency response strategies, using two models of actions that can be called innovative in the context of regular activities of standardization authorities.

Despite the fact that for most standardization authorities, the sale of standards is one of the main sources of financial stability, they have opened free access to the texts of standards that apply to critical goods (the absolute majority of standardization authorities), not only for national users but also for all interested persons, regardless of their location. Given that local and global emergencies may recur in the future, it seems appropriate to keep such "emergency standards" in the public domain and even expand the list to include, for example, standards that apply to life-saving equipment and all personal protective equipment.

10.4 Main Conclusions and Recommendations from the Analysis of TBT Measures During the Covid-19 Pandemic

A study of the use of TBT measures in the response to the pandemic leads to the following conclusions.

Technical measures to regulate the market for goods can be effectively applied in emergency situations, ensuring the timely introduction of necessary mandatory requirements for critical goods, while, at the same time, reducing technical barriers to trade in order to ensure continuity of supply.

In the context of the COVID-19 pandemic, the number of TBT measures aimed at reducing technical barriers to trade related to both conformity assessment procedures (certification and registration) and the temporary suspension of certain product requirements (requirements of regulations or standards) has increased.

As in other areas, TBT notifications issued in March–June 2020 often involve the use of digital (electronic) documents and services; this trend should continue in the future, after the end of the pandemic. Thus, the transition to digital tools in the field of technical regulation is taking place much faster than expected, and this reality shall be taken into account in the practice of technical regulation and standardization, including providing for the organization and financing of the creation of full-fledged digital services and platforms in this area.

There is a tendency to simplify certification procedures (no checks on production and testing of mass-produced products, or certification of only batches of goods) during the pandemic. These approaches should be evaluated from the point of view of analyzing cases of damage (or failures) related to products that have passed the "emergency mode" conformity assessment, but the use of the mechanism for certification of imported goods exclusively under the batch certification scheme seems quite promising.

A significant number of international agreements and collective appeals from countries to reduce restrictions related to TBT measures taken before the onset of the pandemic are shown. In particular, the agreements related to reducing requirements for goods to be delivered in order to preserve supply chains.

During the crisis of the COVID-19 pandemic, the role of standardization increased, primarily national, which was to ensure timely availability of requirements and recommendations related to both the production of critical goods and business sustainability.

Under emergency conditions, traditional standardization tools have changed: there has been widespread digitalization of the process of developing standardization documents; increased availability of standards, widespread use of incomplete consensus documents related to issues of overcoming the pandemic and its consequences, as well as more active participation of standardization bodies in society.

The introduction of open access to texts of international, regional, and national standards during the COVID-19 pandemic, as well as texts of incomplete consensus documents, has proved to be an important tool and can be attributed to "standardization without borders." Given that such global or local emergencies are likely in the future, it seems appropriate to define a set of standards in the field of health, medicine, rescue devices, and personal protective equipment that should be freely available. In addition, national standardization authorities should provide national operators with ways to obtain foreign standards (standardization documents) that are useful for preventing an emergency and overcoming its consequences, including by informing them about their placement on the websites of other standardization authorities.

10.5 Acknowledgements

The authors thank anonymous reviewers for providing critical comments for improving the quality of the chapter.

References

[1] "Notification report - Technical Barriers to Trade," Wto.org. [Online]. Available: http://tbtims.wto.org/en/PredefinedReports/NotificationReport. [Accessed: 01-Oct-2020].

[2] World Trade Organization, "WTO Agreement on Technical Barriers to Trade", WTO, 1995.

[3] Gov.br. [Online]. Available: http://www.inmetro.gov.br/legislacao/rtac/pdf/RTAC002625.pdf. [Accessed: 01-Oct-2020].

[4] Wto.org. [Online]. Available: https://www.wto.org/english/tratop_e/covid19_e/standards_report_e.pdf. [Accessed: 01-Oct-2020].

[5] "WTO members' notifications on COVID-19," Wto.org. [Online]. Available: https://www.wto.org/english/tratop_e/covid19_e/notifications_e.htm. [Accessed: 01-Oct-2020].

[6] "G/TBT/N/IDN/84/add.1 - technical barriers to trade," Wto.org. [Online]. Available: http://tbtims.wto.org/en/ModificationNotifications/View/110409. [Accessed: 01-Oct-2020].

[7] *"CC 817.0 federal act of 20 June 2014 on foodstuffs and utility articles (foodstuffs act, FSA)," Admin.ch. [Online]. Available: https://www.admin.ch/opc/en/classified-compilation/20101912/index.html. [Accessed: 02-Oct-2020].*

[8] "Consultor de Documentos ARCSA," Gob.ec. [Online]. Available: http://permisosfuncionamiento.controlsanitario.gob.ec/consultordocumentos/index.php. [Accessed: 02-Oct-2020].

[9] "WORLD TRADE ORGANIZATION," *WTO.* [Online]. Available: https://www.wto.org/english/tratop_e/covid19_e/notifications_e.htm. [Accessed: 12-Oct-2020].

[10] "WORLD TRADE ORGANIZATION," WTO. [Online]. Available: https://www.wto.org/english/tratop_e/covid19_e/notifications_e.htm. [Accessed: 12-Oct-2020].

[11] "WTO members' notifications on COVID-19," Wto.org. [Online]. Available: https://www.wto.org/english/tratop_e/covid19_e/notifications_e.htm. [Accessed: 12-Oct-2020].

[12] "WTO members' notifications on COVID-19," Wto.org. [Online]. Available: https://www.wto.org/english/tratop_e/covid19_e/notifications_e.htm. [Accessed: 12-Oct-2020].

[13] "WTO members' notifications on COVID-19," Wto.org. [Online]. Available: https://www.wto.org/english/tratop_e/covid19_e/notifications_e.htm. [Accessed: 12-Oct-2020].

[14] Imprensa Nacional, "Imprensa Nacional," Gov.br. [Online]. Available: https://www.in.gov.br/en/web/dou/-/portaria-n-111-de-27-de-marco-de-2020-250196230. [Accessed: 12-Oct-2020].

[15] "WTO members' notifications on COVID-19," Wto.org. [Online]. Available: https://www.wto.org/english/tratop_e/covid19_e/notifications_e.htm. [Accessed: 11-Oct-2020].

[16] Declaration on trade in essential goods for combating the COVID-19 PANDEMIC 15th April 2020. [Online]. Available: http:////perma.cc/WWG4-JRAC. [Accessed: 13-Oct-2020].

[17] *"Declaration of the special ASEAN summit on Coronavirus disease 2019 (COVID-19)," Asean.org, 14-Apr-2020. [Online]. Available: https://asean.org/declaration-special-asean-summit-coronavirus-disease-2019-covid-19/. [Accessed: 12-Oct-2020].*

[18] "Declaration on facilitating the movement of essential goods by the APEC ministers responsible for trade (MRT)," Apec.org. [Online]. Available: https://www.apec.org/Meeting-Papers/Sectoral-Ministerial-Meetings/Trade/2020_MRT/Annex-A. [Accessed: 10-Oct-2020].

[19] "WTO members' notifications on COVID-19," Wto.org. [Online]. Available:https://www.wto.org/english/tratop_e/covid19_e/notifications_e.htm. [Accessed: 13-Oct-2020].

[20] Gov.ru. [Online]. Available: https://roszdravnadzor.gov.ru/i/upload/images/2020/6/9/1591682816.71492-1-34520.pdf. [Accessed: 10-Oct-2020].

[21] "EUR-Lex-32020H0403-EN-EUR-Lex,"Europa.eu.[Online].Available: https://eur-lex.europa.eu/legal-content/EN/TXT/?uri=CELEX%3A32020H0403. [Accessed: 13-Oct-2020].

[22] *Unido.org*, 2020. [Online]. Available: https://www.unido.org/sites/default/files/files/2020-04/Quality%20and%20Standards%20and%20their%20Role%20in%20Responding%20to%20COVID-19.pdf. [Accessed: 07- Aug- 2020].

[23] [8]"COVID-19 Respond: ISO standards, which are publicly available," Iso.org, 2020. [Online]. Available: https://www.iso.org/ru/covid-2.html. [Accessed: 10-Oct-2020].

[24] "COVID-19: ISO members resources" Iso.org, 2020. [Online]. Available: https://www.iso.org/ru/covid19-members. [Accessed: 10-Oct-2020].

[25] "UK's National Standards Body - COVID-19 response," Bsigroup.com. [Online]. Available: https://www.bsigroup.com/en-GB/topics/novel-coronavirus-covid-19/bsi-knowledge-uk-national-standards-body/. [Accessed: 10-Oct-2020].

[26] "Welcome to SAC," Gov.cn. [Online]. Available: http://www.sac.gov.cn/sacen/FAC/RIOCOCAFEPS/. [Accessed: 11-Oct-2020].

[27] "Protective masks: download our reference document for free! - AFNOR Group," Afnor.org, 31-Mar-2020. [Online]. Available: https://www.afnor.org/en/news/protective-masks-download-our-reference-document-for-free/. [Accessed: 10-Oct-2020].

[28] "COVID-19," Nen.nl. [Online]. Available: http://www.nen.nl/covid19. [Accessed: 11-Oct-2020].

[29] Cencenelec.eu. [Online]. Available: https://www.cencenelec.eu/research/CWA/Documents/CWA17553_2020.pdf [Accessed: 10-Oct-2020].

[30] "COVID-19 - Guidelines," Bsigroup.com. [Online]. Available: https://www.bsigroup.com/en-GB/topics/novel-coronavirus-covid-19/covid-19-guidelines/. [Accessed: 11-Oct-2020].

11

Climate Change and COVID-19: An Interplay

Vibhu Jately[1], Jyoti Joshi[2], Rajendra Kumar Jatley[3]

[1]MCAST Energy Research Group, Malta College of Arts, Science and Technology, MCAST, Paola, Malta
[2]Department of Electrical Engineering, College of Technology, G. B. Pant University of Agriculture and Technology, Pantnagar, Uttarakhand 263153, India
[3]Professor Emeritus, Department of Electrical and Computer Engineering, Wollega University, Nekemte, Ethiopia
Corresponding Author: Vibhu Jately, vibhujatley@gmail.com

Abstract

During the last decade, the world has been assessing the adverse effects of climate change on human life, animal life, and plant life. It has been accepted by all Governments and experts that even a 1.5°C–2°C rise in atmospheric temperature will be playing a havoc on our planet. The near unanimity of all countries, which matter, led to the historic Paris agreement for reducing carbon dioxide and other greenhouse gases' emission levels. Efforts by all stakeholders were on to control and contain the disruptive effects of climate change. Amidst these efforts, yet another disruptor, the ongoing pandemic of coronavirus, COVID-19 struck. We are grappling with its ferocity since January 2020. Most countries have imposed restrictions on the movement of personnel and goods to contain the disease and to stop its spread to new areas. This has led to a large-scale suspension of industrial activities across sectors. Most severely hit are persons and organizations associated with transportation, tourism, and hospitality. Trade and commerce have also been adversely affected by the disruption of supply chains. The industrial production is down and a fall in GDPs more severe than the one caused by the recession of 2008 is feared.

In a nutshell, the entire mankind is under economic and social stress due to the suspension of various human activities in its fight against the pandemic. But this suspension of activities has seen some pleasing returns too in the form of cleaner air in most cities across continents. A natural question that comes to one's mind is as to how would the effect of these two disruptors, the ongoing pandemic and man's fight against climate change, affect each other. In this chapter, the short-term and long-term effects of these disruptors are identified, the factors influencing the efforts by various stakeholders are discussed, and pitfalls that may come in the path are highlighted by looking at the similarities and differences in the two disruptors. Finally, a road map for the planners in the light of these discussions is suggested.

11.1 Introduction

A rapid industrialization and urbanization of our society resulted in a rather indiscrete use of our planet's energy resources, mostly fossil fuels, in the twentieth century. This indiscretion in use, coupled with a tendency of caring less for the efficiency of the machinery employed in energy conversion, led to a fast depletion of the limited resources of petroleum, coal, and natural gas. The worse part of this development was the harmful effects it had on the environment in the form of pollution near the industrial hubs and the cumulative degrading of the earth's atmosphere. The long-term adverse effects of the unbridled release of carbon dioxide and other harmful greenhouse gases (GHG) in the process of burning of fossil fuels led to the depletion of the protective ozone layer and a general rise in atmospheric temperature due to the greenhouse effect. Furthermore, the melting of glaciers and polar ice due to a rise in temperature has resulted in a significant rise in sea levels and is feared to submerge the low-level harbors in the near future which could disrupt its local habitat. Environmentalists have been raising these concerns from the last two decades of the twentieth century. From the beginning of this century, these concerns took the form of social and even political movements in some countries. Under pressure from social scientists, environmentalists, public health experts, and other members of civil society, the Governments and planners started discussions at the highest level to find remedies for the impending crisis and to arrive at ways to ensure economic progress and industrialization in a sustainable manner. There was a growing realization of the fact that the effect of good efforts by one geographical entity may be negated by the absence of a similar effort by others. There was a cry for concerted international cooperation in the face of the challenge of climate change.

Extensive negotiations among all major countries culminated in the historical Paris Agreement in December 2015 that brought practically all nations together in order to save the world for our future generations. A concerted effort, both jointly and severally, is envisaged in this agreement to reduce the emissions. This agreement aimed to keep global temperature rise less than 2°C above the pre-industrial levels, as suggested by the Intergovernmental Panel on Climate Change (IPCC). The agreement was formalized in November 2016 and was ratified by 170 countries (including 28 members of the European Union), responsible for 90% of global emissions. The requirements of the agreement included all countries to set "national goals" for limiting the emissions. Appreciating the need to stem the environmental pollution, it created a framework with plans to assess the status at a global level every five years from 2023 onwards. Various committees and groups were formed for monitoring the emission levels and activities for the reduction of these emissions. Among these, the European Union's Joint Research Centre publishes "Science for Policy" reports in the form of Emission Database for Global Atmospheric Research (EDGAR). It compiles global emissions of CO_2 and other greenhouse gases (GHGs) to support transparency and prepare reports to give a periodic status of CO_2 and GHG emissions of all countries. The reports of 2017 [1] and 2018 [2] show encouraging progress in slowing the emissions by EU, USA, China, and Japan, the major contributors to emissions. Some of the developing countries hold the view that it would be unfair to penalize them for any increase in emissions for some time in view of the increasing energy needs of their population. In Section 11.3, a relative share of global emissions of six major players is presented. While efforts in arresting the growth of emissions to combat the disruptor of climate change were on, suddenly another disruptor in the form of the pandemic COVID-19 struck.

In December 2019, China reported the spread of a unique coronavirus in its Wuhan city. The WHO declared it a Public Health Emergency by the end of January. By mid-March 2020, it was categorized as a pandemic of international concern. Since then, the pandemic's ferocity has spread to all continents, has already infected over 27 million people, and has caused 890,000 deaths. In addition, it has taken away livelihoods of hundreds of millions. To control the spread of infection to new areas and to contain the death rates that continue to rise, the movement of personnel is restricted by lockdown and curfew in many countries. Consequently, economic activities are curtailed. Governments have resorted to drastic measures forcing workers, executives, and business houses to hurriedly face the disruption and find ways to tune to emerging realities. In a flash, the COVID-19 has jolted the normal operations and the assumptions across the businesses and industrial world. Presently, the focus of attention is on how to counter the pandemic and reduce the intensity of the recession that will result from the dislocation caused by it.

In the face of this dislocation, one can easily forget that just a few months back, the social and economic impacts of climate change were being debated to look for a collective international response. The governments and the business leaders were busy searching for sustainable solutions for the energy requirements of the society when they suddenly found that the unsustainable disruption due to this pandemic cannot be avoided.

11.2 Comparison OF Two Disruptors, The Climate Risk, and Covid-19

Many are tempted to ask whether the severity and expanse of COVID-19's disruption and the panic it has created will leave the world with a will and wherewithal for any action to face the challenge of climate change and devote efforts in searching sustainable solutions for its energy needs in these trying times. Any rational student would say that the world simply cannot afford to ignore the sustainability aspect in any activity or process we undertake in our quest for the development of society. As soon as a recovery from the pandemic starts, rigorous climate action would have to be resumed looking into its critical importance in the coming decade in meeting the targets set by the Paris agreement. This action would have to be on two fronts. First, to safeguard against the uncertainties of climate change, new infrastructures being created must be climate-resilient. For example, raising height of bridges to account for the sea-level rise or protecting or enhancing the natural drainage systems. Second, for moving toward low-carbon options in the industry and in the generation of electricity. Further, it will also help in creating jobs in a significant number in the near future as efforts are made for increasing economic and environmental resiliency. Currently, central banks are pushing for low rates of interest to help businesses to recover. These rates do not appear to be rising anytime soon. These low rates of interest offer an opportunity to attract investment in sustainable technologies.

To meet this need for sustainability and to make good use of this opportunity, one would need to search for answers to the following:
- We are going through a major disruption over the entire planet under the impact of COVID-19. Do our reactions and responses make us better prepared for the impending disruption that a rise in global temperature would cause?
- What are likely to be the positive implications, and what will be the negative implications of our response to the COVID-19 on our efforts against climate change?
- Are we able to identify and modify actions that we, our governments, and our industrial/business houses are taking to face the pandemic so that the broader cause of sustainability is served?

In the following sections, an attempt is made to provide some initial answers to these questions.

11.2.1 Short-Term and Long-Term Effects of Climate Change and COVID-19

Now that we have been through the current pandemic for about six months, we can enumerate the effects it would have over a short time and over a long time. Similarly, the likely effects of climate change can also be listed.

11.2.2 Short-Term Effects of the Current Pandemic

The effects listed below are mostly temporary in nature:

- decline in carbon emissions;
- increase in medical waste;
- major job losses;
- suspension of scientific, academic, art, cultural, religious, and sports meets.

As one after another country resorted to lockdowns, the suspension of transport activities became a major factor contributing to a temporary decline in carbon emissions. More on it will be discussed in Section 11.4. The current pandemic is causing a large quantity of medical waste. Most countries have their own regulations for storage and safe disposal of this waste. If needed, guidance from WHO [11], Basel Convention [12], and United Nations Environment Programme [14] can be sought. There has been a large-scale loss of jobs due to disruption of industrial activities across sectors. In a survey conducted by International Labour Organisation (ILO) among young people (aged 18–29), 67% workers in low-income group countries reported partial or full decline in working hours compared to 54% in lower middle-income group countries in high-income ones [14]. Academic, scientific, sports, religious, and cultural meets continue to be suspended due to COVID-19.

11.2.3 Long-Term Effects of the Pandemic

Apart from those related to the health of affected persons, some long-term effects, in some regions, can be economic, social, political, and cultural. Enumerated hereunder are the ones that can be affecting our businesses and trade for long:

- substantial reduction in business travel;
- more reliance on video-conferencing;
- some rollback in international trade;
- emphasis on producing goods within the country;
- significant job losses.

11.2.4 Short-Term Effects of Climate Change

For a short time, a rise in global temperature may result in:

- a temporary gain in agriculture output;
- deterioration of plant species;
- uncertain rains and snow.

In some regions, especially in sub-tropical regions, agricultural production may seem to be picking up due to a very temporary respite from cold. Some species may lose the battle for survival. Uncertain rains and snow may also result due to climate change.

11.2.5 Long-Term Effects of Climate Change

As the temperature rises, the melting away of snow at poles, in sea, and the glaciers in the high mountains in Himalayas has major implications for marine life [16]. There are changes in the dissolved oxygen levels in small lakes too by the climate changes that may endanger the dependent flora and fauna [17]. United States Geological Survey website details how the climate change may affect the natural disasters [15]. Some long-term adverse effects of climate change are listed below:

- melting away of polar and sea snow;
- frequent heat-waves;
- heavier precipitation;
- decreased water resources in semi-arid zones;
- natural forests may degenerate into Savannah;
- loss of biodiversity;
- warmer oceans and more tropical storms;
- risk of inland floods;
- water stress in Africa-Asia.

11.2.6 Searching Ways to Mitigate

For finding ways to mitigate the above effects, let us identify the features that are common between the two disruptors and also the features that are different.

11.2.7 Common Features

The most common feature among the risks of global warming and a pandemic is that they shake thoroughly the economy of the world and the social fabric of nations. This shock is physical in nature and similar to a shock associated with a natural calamity or an economic recession. Incidentally, the world has not experienced any physical shock during the last seven decades. Wars after WW2 have been only a local affair, affecting only a very small part of the world. The current pandemic is able to show us what happens when the external forces result in disrupting the supply and demands of commodities. It is also giving us a feel of the effect of unbridled transmission of the disease across borders and it also shows us how the adverse effects are cascaded. A full-fledged climate crisis may also cause similar disruptions of supply and demand. The adverse effects of climate change would also be transmitted across nations and so maybe the cascading mechanism behind their spread.

Some more identical influences of the two disruptors are:

- both tend to dislocate the entire system rather than a part of it;
- past experiences do not give an exact clue of their final effect;
- may lead to a catastrophe if not controlled early;
- both highlight the system's vulnerabilities;
- they affect the lower strata of the society more;
- experts have been warning against both;
- prevention has less cost than bearing the crisis;
- they need action now for a reward much later [3];
- both need international cooperation.

India was quick to realize the last aspect when it offered support to neighboring SAARC countries as early as in March 2020.

11.2.8 Features that Make Them Different

A pandemic has an immediate, direct, and countable effect in the number of deaths, infections, etc. On the other hand, the adverse influences of climate change are gradual and cumulative. For the former, the effects may last weeks, months, or years; for the latter, it may be for years, decades, and centuries. Co-relations may be relatively easy to visualize in the case of a pandemic. Such co-relations may be obscure in the event of a crisis from climate change.

11.2.9 Mitigating the Risk by Avoiding its Multiplication

It does not need rocket science to understand that a rise in global temperature will multiply the risk of a pandemic by causing the spread of vector-borne diseases due to cold regions becoming hotter. Methodical scientific efforts to find direct evidence have confirmed this apprehension [4, 18]. Similarly, if the low altitude areas of the globe are flooded, it may cause animals to flea and may thereby spill over pathogens to humans, manifesting in pandemics like COVID-19. Interestingly, the following actions will help to mitigate the risk of both disruptors:

- use plant proteins in place of animal proteins;
- reduce consumption of energy;
- decrease pollution;
- shorten supply chains.

Some leaders have already been "vocal on local" to lay emphasis on the shortening of supply chains.

We are just seeing how ill-prepared the world has been in the face of the pandemic. Restrictions on the movement of personnel and goods, to contain the infection of coronavirus from spreading to new areas, have led to a major halt of industrial activities in different sectors. Most severely hit are transportation, tourism, and hospitality. This suspension of activities has seen some pleasing returns in the form of cleaner air in most cities across continents [5], albeit, with negatives of downwards slide of economies and loss of millions of jobs.

The issue of achieving environmental and economic sustainability, together, is addressed in Section 11.5. Before that, in Section 11.3, let us look at the resulting trends in response to the international efforts to control the CO_2 and other GHG emissions before the pandemic struck. How the current pandemic is reducing the energy demand will be visited in Section 11.4.

11.3 Trends in CO2 and GHG Emission Levels

The trends of CO_2 and GHG emissions starting from the year 1970 till 2016 have been recorded on the basis of data compiled from all major countries of the world. In Figure 11.1, a plot of total yearly CO_2 emissions of the world is plotted for this period along with those of USA and EU28. The plot shows that for both USA and EU28, the total CO_2 emissions flattened from year 2000 with a small but steady decline till 2016. For Russia, the peak occurred even earlier in 1992; emissions declined in the nineties and are now maintained nearly constant. However, the world's total emissions have been rising due to increased emissions associated with the rapid development of China. Interestingly, even in the case of China, it plateaued in about 2015 and held there. From the plots, it can also be seen that the bulk of the emissions are from transport and from non-combustion sectors.

Per capita yearly CO_2 emissions from major contributors and from the entire world are shown in Figure 11.2. It can be seen that there has been a steady decline in per capita CO_2 emissions of USA, EU28, and Russia,

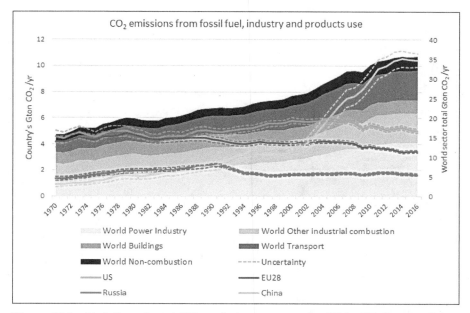

Figure 11.1 Variation of total CO_2 emissions per year by USA, EU, Russia, China, and world [2].

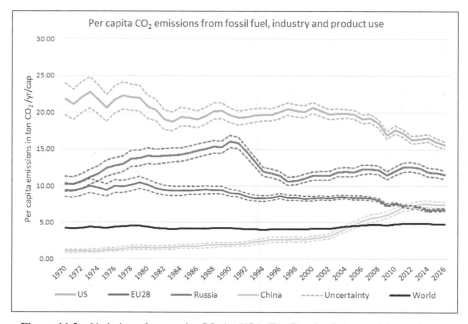

Figure 11.2 Variation of per capita CO_2 by USA, EU, Russia, China, and the world [2].

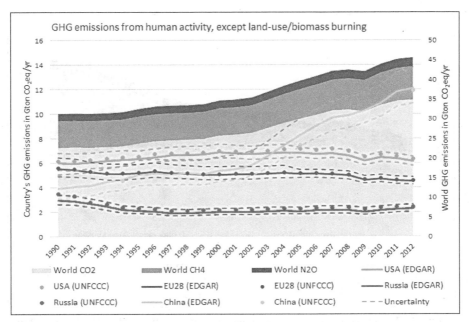

Figure 11.3   Variation of total greenhouse gas emissions per year by USA, EU, Russia, China, and the world [1].

whereas, for China, it has been rising. In the year 2014, China's per capita emissions slightly overtook those of EU28. For the world, there has been a slight but a steady increase in per capita emissions due to rising contributions from developing economies like India.

Figures 11.3 and 11.4 show the total yearly GHG emissions and per capita yearly GHG emissions, respectively. These figures show trends of GHG emissions for the world, USA, EU28, Russia, and China. Here also, it can be seen that positive effects of fall in total emissions of USA, EU28, and Russia are more or less negated by the increased emissions of China and other developing countries with the net result that for the entire world, the emission levels continue to increase at a small rate.

A summary of CO_2 emissions by six big emitters is given in Table 11.1.

From the trends of CO_2 and GHG emissions since 1970, it can be seen that much is still required to be done and done expeditiously. Action on the following lines can be suggested:

- There is an urgent need for developing countries to decouple these emissions from GDP growth. Retrofitting of existing less-efficient

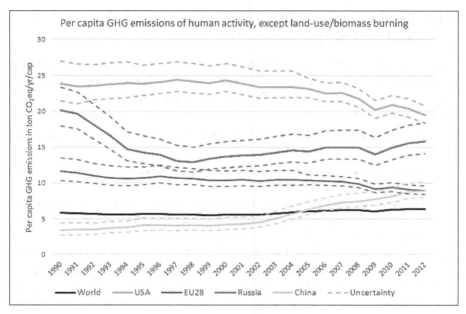

Figure 11.4 Variation of per capita greenhouse gas emissions by USA, EU, Russia, China, and the world [1].

Table 11.1 Percent share of global, total emissions (Gt), per capita emissions (t/cap/yr) and trends of emission by six biggest emitters of CO_2. (All data courtesy JRC, EDGAR, European Union.)

Emissions/Major Emitter	EU	China	USA	India	Russia	Japan
Percentage of global emissions (2016)	9.6	29.2	14	7.1	4.7	3.5
Percentage of global emissions (2017)	9.43	29.3	13.89	6.74	4.85	3.51
Change (%) in 2016 over that in 2015	0.2	−0.3	−2	4.7	−2.1	−1.2
Per capita emissions in t/cap/yr in 2017	7.0	7.7	14.82	1.8	12.3	10.4
Total emissions in 2017 (Gt)	3.5	10.9	5.1	2.5	1.8	1.3

systems in industry and transportation and increasing share of renewables in the power sector can help in reducing emissions.

- Developed countries should quickly bring down per capita emission levels, for which per capita consumption of energy may also have to be, perhaps, brought down.
- The developed world should be more forthcoming in technically supporting and funding the green technology initiatives in developing countries in Africa, Latin America, and Asia.

11.4 Effect of Covid-19 on Emission Levels and on Energy Demand

Data on the drop in energy demand and in CO_2 emissions through mid-April from 30 countries, bulk from Europe and North America, when lockdown instructions were in force there, were collected by International Energy Agency (IEA). These data were used to project these drops for the entire year [6]. Their key findings are:

- The energy demand in April fell by 18%–25% in these regions depending upon how strongly the lockdowns were imposed.
- There was a drop of 3.8% in the energy demand for the entire world in the first three months of 2020.
- Demand for coal fell by 8% in January–March 2020, from what it was in the corresponding three months last year. Shutting down of coal-fired plants in China could have contributed to this fall.
- The demand for oil fell by 5% due to restrictions on transport, on the sea, on the surface, and in the air in the same period of three months.
- Demand for gas fell only marginally, though the renewal energy saw a small rise.
- Full lockdown periods saw a drop of 20% in electricity demand.

Second quarter, April–June, saw major restrictions in India with implications for a further drop in coal and electricity demands. Energy demand is projected to fall by 6% in the full year; largest in the last seven decades and seven times more severe of the fall that occurred in 2008 due to recession. In this fall, major contributions come from oil – 9% and coal – 8%. Gas and nuclear may see a marginal drop and renewables a small gain.

In 2020, electricity demand may see a drop of about 5% (up to 10% in some areas), as shown in Figure 11.5.

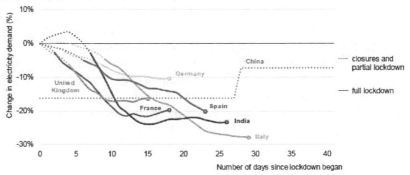

Figure 11.5   Projected fall in CO_2 emissions in 2020 [6].

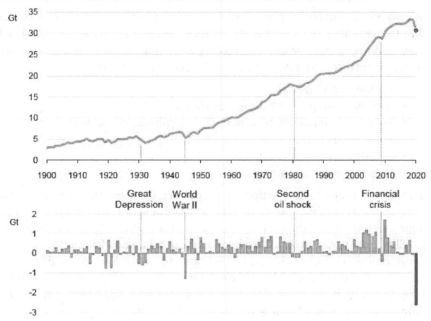

Figure 11.6 Projected drop in electricity demand [6].

Overall, an 8% fall in CO_2 emissions, i.e., 2.6 Gt may result in this year. This fall shall take the emissions for the world close to a level of 2010, as shown in Figure 11.6.

11.5 How to Move Forward?

Half a year has gone by for the world in its struggle against the COVID-19. The number of cases infected and deaths reported in different regions/countries have been grossly varying. So have been the trajectories of the curves of case-count and number of deaths, with peaks occurring on different dates. There is still a lurking fear of a virus's mutation for the worse or a second surge once the restrictions are lifted in an effort for recovery. Yet, based on the response and reactions of all stakeholders, we are in a position to segregate the actions that can help us in the climate effort from those that can blur our focus.

11.5.1 Responses Helpful in Saving the Environment

We all are a witness to a sudden improvement in the air quality in our cities due to a number of contributing factors. Several of these, as listed below, may become a part of our changed routine and, in turn, may augment the collective effort of mankind in reducing the harmful emissions.

- working from home;
- increased adoption of digital dealings;
- reduced demand for transport with implications of reduction of Scope 3 emissions [7];
- shortening of supply chains;
- reliance on local produce.

The above factors have helped in reducing the pollution levels by doing away with avoidable travel. In addition to these, of late, there is a greater appreciation for building a pool of experts for addressing systemic problems [8]. The governments have also been realizing the need for intergovernmental cooperation in facing the pandemic and for coordinating actions against climate change. In addition, lower interest rates offer an opportunity for investment in green technologies and resilient infrastructures. That may also help in the creation of new jobs.

11.5.2 Pitfalls in Road that Can Blur the Focus of Stakeholders for Reducing Emissions

As governments, organizations, and individuals struggle to limit the socio-economic effects of COVID-19, and as the priority is on saving lives by investing quickly and heavily on augmenting the public health infrastructures, there is a chance of various stakeholders losing the focus of their commitment to achieve sustainable development. Some of the pitfalls in this journey are as follows.

- Foremost among these are financial:
  - Empty coffers can impede the switchover to green technologies by the governments, as the bulk of the budget would be diverted to support the medical facilities to limit the death count due to the pandemic and to provide health care to the most vulnerable.
  - Jobs are lost all over the world and livelihoods in the informal sector are threatened. Providing social security has become a priority. Social welfare schemes and public distribution system, especially in the developing countries, need much larger outlays to feed the hungry, leaving little money with the administration for climate action.

○ Supply chains are disrupted. This has left producers in distress due to a fall in demand. The consumers' pockets are empty. Even those who have some spare money are tempted to defer buying non-essential goods to better times after the pandemic recedes. Under such conditions, manufacturers may find it difficult to find cash for a shift to greener technology.

○ Financial stress may result in the end of or lowering of the subsidies to prosumers in setting up PV and wind-based generating units. Other regulatory exemptions like lower taxes may also go away.

• Risk of adoption of half-baked technologies:

○ There is a danger of increased CO_2 emissions as some countries try to control the pollution in urban areas by putting devices to control SO_2 levels. Sulfur dioxide in the air is known to be a lot more harmful than CO_2 for humans. Some companies are offering devices that are claimed to capture SO_2 from the flue gases, especially those from coal-fired thermal power plants. But these devices boost total CO_2 emissions [20].

○ The installation of carbon-capture systems is another yet-to-be perfected technology. What to do with sequestered CO_2 is yet not clear [21]. Unless a proper beneficial, economical, and safe use of the captured CO_2 is identified, a large-scale use of the concept remains a pipedream.

• Fall in international hand-holding:

○ In times of crisis, political leaders are often tempted to dispense an excessive dose of nationalism, which may lead to interstate rivalries halting the international cooperation that is so necessary for climate action.

○ Even before COVID-19 struck, USA, one of the largest carbon emitters, was threatening to come out of the Paris Climate Agreement. If the threat were really carried out, the situation may spiral out with others also going their own way.

11.5.3 Road Map for the Planners

We normally think of the role of planners only among the governments. As the onslaught of COVID-19 has shown that apart from the governments, actions are needed at many levels, for example, by industrial manufacturers, the stockists of products, the distribution agencies and by the consumers at large to mitigate the adverse effects of the disruptors. At all levels, people, though still learning, have become a little wiser and have realized that such disruptions can recur in the future. With this realization, a need for planning for coming disruption is being thought of. Challenge of a disruption by global warming that the world sees at the horizon calls for planning by stakeholders at all levels.

Planners in the governments may consider actions on the following lines.

- Invest in creating expertise that can prepare reliable models on the basis of historical trends on the likely gross emissions in the near future, the adverse effects these emissions will have on the climate, and how risky these changes in climate would be in areas under their domains. The modeling should also cover the financial implications of desired improvements in infrastructure in the affected areas so that these infrastructures are more resilient in mitigating the ill effects of climate change on the people.

- A portion of the national budget should be earmarked for funding a shift toward environment-friendly systems. New buildings should be designed to be resilient to climate changes and the existing ones may consider some retro-fittings. The promotion of solar PV and wind-based systems should be a priority. This would entail some changes in the grid structure, for which funds should be apportioned. Sustainability should become the key concept followed and ensured in heavy industries; decarbonizing efforts in heavy industries should be encouraged by incentives. Money spent in these efforts will result in creating new jobs and will simultaneously slow the rate of rise in global temperatures.

- Subsidy regimes on using energy sources that are harmful to the environment must be done away immediately. Some countries are subsidizing the use of coal in industries and in power generation simply because they have enough reserves for it. They should seriously rethink this approach.

- The need for international cooperation in arriving at sustainable solutions should be understood by all participants of the Paris Climate Agreement. Individual efforts may not be as effective as a collective one globally.

- There is an urgent need to put in place strong incentives for the adoption of green technology for the industry and individuals. Similarly, strong penalties should also be imposed on manufacturers using inefficient and high emission processes. Imposing carbon taxes on their products, in many cases, shall prove to be a strong discouragement for continuing old technologies.

- Plan a complete paradigm shift in the power sector. Let us not forget that coal-fired thermal power stations alone account for nearly 30% CO_2 generation. Efforts are on to develop viable technology for carbon capture [22]. Similarly, some techniques for safe removal of SO2 are also being evaluated [19]. Till some commercially successful technologies are achieved, a hold on coal-based electricity generation or at least a reduction in its use should be aimed. Planners must push for renewables and for distributed generation to enhance reliability and reduce emissions.

- Upgrade pollution/emission standards from the vehicle transport sector, which is one of the major polluters [9]. This sector accounts for nearly 12% of the total emissions in the European Union. Incentives should be provided to users for switching over to electric vehicles (EVs). For the adoption of EVs, the creation of infrastructure for charging stations (for batteries) must be planned expeditiously. Even the parking lots with bidirectional charging facilities should be quickly put in place. The latter may help in supporting the micro-grids to tide-over the problems in the event of any failure on the main grid, and with appropriate control assist in arriving at optimal day-ahead schedule for charging [23].

Business houses may consider action on the following lines.

- Use the current pause in activities to phase out obsolete high emission machinery and replace these with more efficient and environment-friendly modern equipment. This would, in general, result in reducing recurring expenditure on operation and maintenance, apart from participating in the drive to save the planet for posterity.
- In view of COVID-19 and associated restrictions, all major firms are already constrained to resort to innovative measures like digitalizing of their dealings, work from home, use of shorter supply chains, making do with fewer staff, staggering operating hours, holding video conferences, reducing travel in connection with business, etc. They may find some of these innovations worth continuing even after the pandemic recedes as a matter of economics and efficiency.

Besides the points mentioned above, the business houses must prepare themselves to face the physical shocks of two or more disruptors simultaneously. It must not be difficult to imagine a hurricane, a cyclone, an earthquake, a tsunami, a major fire, or a flood occurring when a pandemic is sweeping an area. They must workout models to forecast the probability of their occurring simultaneously and forecast the effects of their simultaneous occurrence on their businesses. Finally, they should build resilience in their procedures, processes and systems so that the adverse effects of such an occurrence can be controlled.

Current turmoil offers us all an occasion for introspection.

- It is time to have an appreciation and to build awareness about how damaging and how long-lasting the effects of disruption caused by a rise in global temperature can be. All must realize that lack of resilience and redundancy can result in a catastrophe of enormous proportions.
- During the last few months, most of us have changed our lifestyles, though perforce, in a way that our collective demands on the environment are much less. Less of indulgence and a controlled temptation to flaunt improve not only the individual character, it also augments our savings, uplifts us spiritually, and satisfies our innate feelings for having contributed to the noble cause of caring for the Mother Nature.
- All should seriously consider discontinuing the use of a personal car to the workplace. In India, buses constitute only about 1% of total automobiles, whereas a bus can carry passengers that are equivalent to numbers that 30 cars can carry [24]. These ratios of numbers of buses to cars in other developing countries would not be very different. If people can go to offices, factories, and other places in buses rather than in cars, the savings on fuel, reduction in emissions, decongestion of road traffic, and reduction in requirements of parking areas in workplaces can all simultaneously be achieved.

We have just been noticing how unprepared the world is in facing the disruption caused by COVID-19. We must prepare for another physical disruptor due to global warming. Coming years are going to be crucial for all of us in our fight against climate change. We must understand that if the emission levels do not fall quickly, a disaster may be lurking around the corner. Global emissions crossed 32 Gt/year in 2019. A level of 40–50 Gt shall result in about 2°C rise in average global temperature. At the current rate of rise in emissions, this may happen in about 15–20 years. With little carbon capacity left with us, the task of reducing emissions will become more daunting every year. Avoiding catastrophic climate change is a collective responsibility of all governments, business houses, and the public at large. The only way to survive is by using the positive implications of COVID-19's disruption as an opportunity and to help in achieving net-zero emissions by 2050 [10].

11.6 Conclusion

The world continues to reel under the onslaught of pandemic COVID-19 showing how unprepared we have been in facing this disruptor, thereby serving as a wakeup call to all stakeholders for the challenge of the other disruptor of climate change that this planet may have to face soon. In this chapter, the short-term and the long-term effects of these two disruptors have been identified; so have been their similarities and dissimilarities for mitigating their adverse effects. The trends of carbon emissions of major players in the light of international efforts to control these carbon emissions show the gaps that call for immediate cooperative follow-up. A road map by way of recommendations for planners, business houses, and public at large for coordinating the efforts for fight against the climate change has been presented.

11.7 Acknowledgements

Material compiled here relies heavily on reports from McKinsey Quarterly, reports of JRC (EDGAR), European Union, IEA, and World Economic Forum. Authors are grateful to these organizations for keeping these invaluable knowledge resources in the open domain.

References

[1] Janssens-Maenhout et al, JRC for Science Report (EDGAR), 2017.

[2] Muntean, M. et al, JRC for Science Report (EDGAR), 2018.

[3] Carney, Mark, 'Breaking the Tragedy of the Horizon – climate change and financial stability' Sept., 2015

[4] Jordon Rob, 'How does climate change affect disease', Stanford's Earth Matters, Mag. March, 2019

[5] Corinne Le Quere et al, 'Temporary reduction in daily CO2 emissions during COVID-19 forced confinement', Nature Climate Change, Vol.10, July 2020.

[6] 'Global Energy Review', International Energy Agency. Report April 2020.

[7] FAQ, 'Greenhouse Gas Protocol', World Resources Institute, www. GHGPROTOCOL.ORG

[8] Dickon Pinner, Matt Rogers, and Hamid Samandari 'Addressing climate change in a post-pandemic world' McKinsey Quarterly Report, April 2020.

[9] European Parliament and the Council Regulation (EU) 2019/631 on CO2 emission performance standards for new passenger cars and new vans for 2025 and 2030.

[10] Maria Mendiluce, 'How to build back better after COVID-19', We Mean Business, World Economic Forum, April 2020, www.weforum.org

[11] World Health Organization (WHO). Rolling updates on coronavirus disease (COVID-19). https://www.who.int/emergencies/diseases/novel-coronavirus-2019/events-as-they-happen.

[12] Secretariat of the Basel Convention. 2020. Waste management an essential public service in the fight to beat COVID-19. Basel Convention. 20 March. http://www.basel.int/Implementation/PublicAwareness/PressReleases/WastemanagementandCOVID19/tabid/8376/Default.aspx.

[13] United Nations Environment Programme (UNEP). 2020. Waste management and essential public service in the fight to beat COVID-19. 24 March. https://www.unenvironment.org/news-and-stories/press-release/waste-management-essential-public-service-fight-beat-covid-19.

[14] 'Youth and COVID-19', Impacts on Jons, Education, Rights and Mental Well Being, Survey Report 2020, ISBN 9789220328606 (web pdf)

[15] https://www.usgs.gov/faqs/what-are-long-term-effects-climate-change-1?qt-news_science_products=0#qt-news_science_products

[16] Ronald E. Thresher et al, 'Depth-mediated reversal of the effects of climate change on long-term growth rates of exploited marine fish', Proceedings of National Academy of Sciences of the United States of America, PNAS, 2007.

[17] Brian Foley et al, 'Long-term changes in oxygen depletion in a small Temperate Lake: Effects of Climate Change and Eutrophication, Freshwater Biology, Wiley online Library, 2012.

[18] Kovats, R. S., et al, 'Early Effects of Climate Change: Do they include changes in vector-borne disease', Philosophical Transactions of Royal Society B, Biological Sciences, July 2001, https://doi.org/10.1098/rstb.2001.0894

[19] Roy, Papiya and Sardar, Arghya, 'SO2 emission Control and finding a way out to produce Sulphuric Acid from Industrial SO2', Journal of Chemical Engineering and Process Technology, 2015.

[20] 'Govt's pollution control devices for coal-fired power plants can also boost CO2 emissions': Bloomberg Report, Economic Times, 19 December, 2019.

[21] Eldardiry, Hisham and Habib, Emad, 'Carbon Capture and Sequestration in Power Generation: review of impacts and opportunities for water sustainability', 'Energy, Sustainability and Society, (2018) 8:6, Springer

[22] sanders, Robert, 'New technique to capyure CO2 could reduce power plant greenhouse gases', Berkeley News, 23 July, 2020.

[23] Cai, Hui et al, 'Day-ahead optimal charging/discharging scheduling for electric vehicles in microgrids', Protection and Control of Modern Contro Systems, (2018) 3:9, Springer open.

[24] Usman Nasim and Vivek Chattopadhyay, 'Indian roads belong to motorised vehicles, not cyclists or pedestrians; Vehicular emissions cannot be controlled without strengthening public transport system', Down to Earth, 06 November, 2018.

12

COVID-19 Pandemic: A New Era in Higher Education

Sriperumbuduru Srilaya, Sirisha Velampalli

CR Rao AIMSCS, University of Hyderabad, Gachibowli, Telangana 500046, India
Corresponding Author: Sriperumbuduru Srilaya, srilaya789@gmail.com

Abstract

The novel coronavirus has influenced emotional well-being of citizens, agriculture sector, employment, and energy throughout the world as well as other areas of the economy. Coronavirus has remarkably interrupted the higher education sector which is the main deciding factor for economic future of the country. The novel coronavirus has bashed our education system to the extreme level; we can expect a different kind of educational model to emerge after COVID-19. Higher education institutions (HEIs) such as colleges, universities, and tertiary education institutions were in huge loss due to prevailing situations. According to UNESCO, over 1.5 billion students in 165 countries are in huge loss as they cannot attend school. The Government of India conducted survey according to Ministry of Human Resource Development (MHRD) on higher education and observed that there are 993 universities, 39,931 colleges, and 10,725 stand-alone institutions listed on their portal, which contribute to education. This forced global communities to enhance new techniques for learning which also includes both distance and virtual learning. Apparently, it is the first time in India to experiment with education system merging from classrooms with online learning. Educating the students simultaneously about the emotional intelligence to control their fears and anxiety can render their mental stability, and education system will transform in the best way if there is psychological support for students .This work mainly focuses on the changes that need to be incorporated to get over the huge loss for the higher education system due to COVID-19.

12.1 Introduction

On 11 March 2020, COVID-19 has been declared a pandemic by World Health Organization (WHO) and it has affected more than 10 million lives worldwide. The first affected case was detected on 30 January 2020 in the state of Kerala. The first death in India was on 12 March 2020 and the nation declared Janta Curfew on 22 March 2020.The first phase of lockdown was announced by the Prime Minister on 25 March 2020 for 21 days. By observing the changes that the country has undergone due to the prevailing situation, the Government has declared to extend the lockdown till 30 June 2020. Throughout the complete phase of lockdown from lockdown 1.0 to 5.0, the country has never got any alleviation to start the educational institutions. Thus, the pandemic has crucially affected the education sector. Till the first week of 2020, the report of UNESCO clearly explains that COVID-19 has affected nearly 70% of total world student population. COVID-19 pandemic has influenced 1.5 billion students across the globe by school and university termination. In India, more than 35 crore students were affected due to various limitations and nationwide lockdown. Most universities and educational institutions have made the maximum attempt to control the spread of pandemic and make the educational institutions run as usual, but they failed; so, further, they facilitated by digitalizing the sector and provided an opportunity for students using all digital modes of learning [1]. This pandemic has imposed universities to bring their courses online. However, this is one step along a new education system and we can look forward to a new model popping up once COVID-19 has passed [3]. The first things that were affected by these closures are the structure of teaching and learning methodologies and only few private organizations can acquire teaching methods [7]. The students along with lack of facilities for learning do not even have access for healthy meals and are subjected to social and economic distress [8]. Furthermore, there is even struggle for the teachers to maintain the same amount of depth with students as they could perform in classroom learning [3]. They need to find a solution to establish the same quality in the education they are providing right now. Online education platforms such as Master Class, Udemy, Edx.org, Udacity.com, Coursera.org, Futurelearn. com, etc., play a crucial role by digitalizing the market by providing the best content. At present, video-conferencing apps such as Zoom, WebEx, etc., are providing universities a lifeline.

The main objectives of this work are: (i) modification in higher education system by considering affordability of parents and students; (ii) major improvement for modulating and regulating the quality of higher education during crisis; (iii) continuity in dealing with changes toward research and teaching; (iv) long-term changes caused in learning and teaching due to sudden shift in new resource development; (v) changes required to maintain employability of students who are already placed or likely to be placed.

12.2 Covid-19 Impact on Higher Education

COVID-19 has significantly affected the total education system in India as well; some of the influenced areas due to the pandemic are listed below.

12.2.1 All Educational Activities are Disrupted

COVID-19 has implemented lockdown including education sector. Activities such as examination, admission and competitive exams, and entrance tests conducted by various national and international institutions are all postponed. This perhaps created a great challenge in the career of the student. The primary goal of the educational institutions is to conduct the classes without the physical presence of the faculty and the solution for this is digital mode of learning [2]. This has been implemented within a short period of time by the HEI by the end of the first phase of lockdown. COVID-19 has made HEI technologically savvy for the adaption of digital technologies to deliver education [3]. HEI conducted different programs, meeting, and video conferencing sessions through Zoom, Google Meet, WebEx, Skype, YouTube Live, etc., and this provides services to the students. Institutions have also started internship reports and projects through email during lockdown.

12.2.2 Turndown in Employment Opportunities

Many of the entrance exam recruitments got canceled, which created a negative impact on life of students. Many Indians also lost their jobs abroad. In India, there is no recruitment in Government sector and fresh candidates are scared of their employment opportunities [13]. The passed out students may not get the job due to the prevailing situation. All these factors increase the unemployment rate in country.

12.2.3 Impact on Academic Research and Professional Development

COVID-19 has bought a negative impact on researchers as it is impossible for them to work together nationally and internationally as there is no availability of transport facilities [13]. Some joint research work is complicated to implement and scientific research work and testing could not be conducted. Advantage that happened in this lockdown is that academicians got much time to improve their knowledge in a theoretical way and they got enlightened with technological methods and improved their research [4]. They shared their expertise using webinars and e-conferences. They strengthened their technical skill and got scope to publish articles in conferences journals and publish books in free time.

12.2.4 Attendance of Students May Slow Down

Most of the parents resist sending their students back to schools and colleges suddenly after lockdown. Some poor family parents who have lost their livelihood due to pandemic cannot afford the expenditure.

12.2.5 National and International Student Mobility for Higher Study May Be Reduced

The international education has been affected by the crisis. Many international conferences have been canceled and turned into webinars. Social distancing will continue for the next few years and may affect the on-campus face-to-face teaching and learning [7]. Most of the parents will prefer to find the workable alternatives within their home and will restrict less travel within the country. Student safety and well-being issues are important factors for students and their parents to study in international institutions for higher studies.

12.3 Challenges of India for Higher Education During Covid-19

A pandemic like COVID-19 occurs once in several countries. Unfortunately, when it occurs, it disrupts the life of human extremely. Furthermore, it changes social, economic, and political activities in different parts of the world. It stupefies people and leaves them helpless. Nowadays, humanity is seized unaware and disarranging for finding successful survival and growth strategies. The future of human relay depends largely on high grade education and health. Health provides the elixir to the body, whereas education is spirit of mind. Several economic and social activities have suffered from COVID-19. Teaching and education activities have been obstructed drastically from COVID-19. Admission processes are detained. Unfortunately, even placements of students have been revoked creating distress and dismay for youth [8]. Higher education has been contrived by corona in three ways. First, higher education has noticed certainty of online education.

Earlier classroom learning and online learning are hardly mixed. Online learning has been an irreplaceable tool that stays in higher education forever. Online teaching and on-campus teaching are integrating as if it did not happen before. Many efforts are required to make this blending more effective. Information technologies and stronger communication are required and teachers need to be trained to enhance the online teaching–learning process. Students have to be methodical self-learners and the government is required to change the regulations to ease higher education institutions (HEIs) in building up their impression [6].

Moreover, HEIs have to develop a novel approach to allocate students into small groups and engross them in various academic processes. Thus, this is leading toward a contradictory situation. Students will attend online lectures in large groups and they also do other activities such as career development lab work. It indicates that HEIs have to invest enormously in equipment, facilities, and building of digital learning materials [6]. In future, learning will not be counted in terms of class credits and clock hours that students spend in classroom but will be counted in terms of learning credits and the hour students spend in learning units. Assessments and examinations have to be defined and used as learning. Learning is assured perfectly only if there is a greater need for more assessments [12]. At same time, there should be no distress in learning; it should go in a smooth and straightforward way [10]. Sharpness and sense making will aid academics in developing the new paradigm of higher education; by this, a new prototype will emerge slowly and consistently. It also involves various HEIs that must participate together, design learning methods, redefine and develop new modes of learning, and plan and rework on assessment techniques [12].

12.3.1 Virtual Platforms in Higher Education at Times of COVID-19

The HEIs have adopted major strategies to overcome the loss during the pandemic. The Government of India took all the measures to control the spread of pandemic. There were many platforms launched by University Grants Commission (UGC) and Ministry of Human Resource Development

(MHRD) by providing various virtual platforms with e-books and online depositaries and other online learning methods. The digitalized technology of MHRD and UGC during the pandemic are:

i. SAKSHAT (https://sakshat.ac.in/)
It is an education portal for addressing learning related to scholar's teachers and students. This portal provides press releases and latest news related to MHRD.

ii. NEAT (https://neat.aicte-india.org/)
This portal is used for learning new technologies through public–private partnership between the government and education technology companies in India.

iii. Shodhganga (https://shodhganga.inflibnet.ac.in/)
It is a portal for research students to publish their thesis available to the entire scholarly community. This portal has ability to index store and capture.

iv. VIDWAN (https://vidwan.inflibnet.ac.in/)
This is a national research network database; it contains profiles of research/scientists working at academic institutions and other Research & Development organizations in India.

v. e-ShodhSindhu (https://ess.inflibnet.ac.in/)
It is a collection of e-books, e-journals, and e-journal archives on long-term access basis. It has 36,000+ e-books and 10,000+ e-journals. It provides access to factual databases to academic institutions at a low rate of subscription.

vi. Virtual labs (http://www.vlab.co.in/)
It has 100 virtual labs and consists of 800+ web-enabled experiments. It designed curriculum-based experiments for remote operation. It provides remote access for labs in various organizations in disciplines of science and engineering. These labs are provision for research scholars, postgraduates, and undergraduates.

vii. FOSSEE (https://fossee.in/)
It is an open-source software for education as well as for professional use.

viii. e-Yantra (https://www.e-yantra.org/)
It has 380 labs which can be used by 2800+ colleges. This provides hands-on experience on embedded systems.

ix. National Digital Library of India (NDLI) (https://ndl.iitkgp.ac.in/)
It is a repository of e-content on multiple disciplines professional library users, teachers, students, and researchers. It is developed by

IIT Kharagpur. It is developed to help students opt for entrance and competitive exams. It is also available to access through mobile apps.

x. e-Pathya (Offline Access)

It is a software that maintains course and content package that helps students pursuing (PG level) education at campus mode and distance learning mode. It also facilitates offline access.

xi. Swayam provides Massive Open Online Courses (MOOCs) with 140 universities; it provides educational programs through 32 DTH channels. e-PG Pathshala (https://epgp.inflibnet.ac.in/) postgraduate students can access this platform for study materials, online courses, and e-books.

xii. e-Adhyayan (e-Books)

It provides 800+ e-books and these books are obtained from e-PG Pathshala; these are books related to postgraduate courses and provide video content playlist.

xiii. e-GyanKosh (http://egyankosh.ac.in/)

A national digital repository developed by open and distance institutions in India. This repository is used to store digital learning resources. Content in e-GyanKosh has all rights reserved by Indira Gandhi National Open University (IGNOU).

12.4 Challenges Undertaken for Digitalizing Sector in Higher Education

12.4.1 Resource and Internet Connectivity

Majority of population in India still does not have Internet connection; so a huge number of people in India and in remote areas are ignorant in the field of digital technology [15]. One of the major requirements for digital education is Internet connectivity in rural areas and few parts of urban areas.

12.4.2 Shortage of Trained Teachers

There is a lack of trained teachers in digitalized sector [11]. In rural areas there are few educational institutions, schools and professors, teachers in that institutions are very much inactive in using digital tools for conducting classes. But with this pandemic, many of them have adopted digital learning [11].

12.4.3 Content-Related Challenges

A major barrier for evolution of digital education in India is language, as many distinct languages are being spoken across the nation, and, sometimes, it becomes difficult for agencies to load all the regional languages. Most of the Ed-Tech content is in English, which can be accessed by all.

12.4.4 Poor Maintenance and Upgradation of Digital Equipment

There are a lot of financial issues in rural areas, and this leads to poor maintenance and upgradation of digital equipment [11].

12.4.5 Inadequate Funds

Digital sector requires productive as well as well-organized utilization of relevant and current software and hardware technology accessible in digital sector. In developing nations like India, huge funds and infrastructure are required to blend digital technology implementation into the education system.

12.5 Post-Covid-19 (Digitalization: A New Phenomenon in Higher Education)

A rise in the health crisis and national lockdown were making noticeable changes in the education of students as well, by shutting down the universities and their syllabi stranded. The universities have decided to digitalize the sector by reinventing the new technology and making new strategic approach expand in times of disaster [17]. The COVID-19 era in India brings a huge contrast between reinvention in the education system by allowing the HEIs to embrace virtual learning and permeate an online study culture. Pandemic is navigating sector ahead with advancements and technical innovations.

COVID-19 has bought a great opportunity for the educational sector from primary schools to universities and high learning institutions to shift the education system in a digitalized way [16]. The revolution has already started for digital transformation [17] and digital India has arrived, be it in trade, service governance commerce and even in education, digital transformation is clearly visible. India is looking for effective interventions to impact the behavior of people and to build responsible citizenship and a better community that impacts basics on human excellence. Importance of integrity is better understood in e-commerce and e-business as customer satisfaction is very much important and trust of stake holders is much important to enhance business considerations and prosper in future. We need to increase integrity and compassion as this adds core values to build an education system effectively and it will transform human into humanity and professional integrity and humility into core human instinct. The digital transformation in education should not let go an opportunity of merging integrity in digital learning [18]. At this challenging time everyone should utilize this opportunity to civilize the digitalization of the sector.

12.5.1 Urge for Distance Learning and Online Learning May Grow

There is no loss in switch to the online education because their progress is tracked by timely evaluation. Probably, this is the first time in India to demonstrate with education system and build the transition to the world by incorporating classrooms with online learning. Online learning enables students to learn creatively by forming a collaborative approach between education and technology [18]. Universities are engaging students to learn by choice and not just by physical appearance by innovative technology.

Moreover, adopting AI-based techniques in the education systems enables the students to learn more effectively. Virtual internships, being one of the major opportunities in this pandemic, help students to go beyond their curriculum and enable them to master feasibility with respect to their professions [14]. Universities are motivating students to notice the ongoing situation and making them to automate themselves so that this will further allow them to digitalize in their respective fields and along with preparing for this kind of situations so that this practice will inculcate confidence rather than panic. The universities are sensitizing their faculty to address the situation carefully by teaching much more than syllabus apart from virtual and interactive learning [18]. Educating the students about present state of confusion and panic is not only preserving their sanity but also motivating them to be in such distress and crisis. The education system will be transformed to good only if we provide psychological support to students; this, in turn, improves their emotional intelligence.

12.5.2 Blending Teaching and Learning with Technology

Education plays a key role in the evolution of stable and advanced society, polishing skills of human and developing the personality of individuals. By using technology as a catalyst, there is a huge difference in moving knowledge transfer model to self-directed active model [16]. Due to the corona virus, there is a stream of flow toward digitalization noticed in higher education. Technology is engraving its authority day-by-day.

The mode of teaching higher education has changed extremely in the last few years. There are some old guards who stay even with chalk-talk technology, but the corona crisis has forced to use technology in education [6]. Ample amount of information on any subject is available on YouTube, Wikipedia, Google, etc. In India, many universities have opened campuses abroad and have signed MOU-Memorandum of Understanding with some foreign universities to offer online education. By 2040, India will be one among the youngest nations in the world. Over 150 million students, one in every four, will be a student of India [14].

Over the last two decades, there is a remarkable change in higher education landscape [6]. Widespread of access is created to low-cost high-quality university education for all levels of students.

12.5.3 New Design in Assessment System

AI uses digital platform in handling examination and evaluation system to reduce the burden of examiner. It also helps teachers in monitoring the performance of students easily preparing mark-sheets, assessment, evaluation, etc. [12]. A quality assurance mechanism and quality benchmark for online learning offered by HEIs and well as e-learning platforms is uniformly important. There are many e-learning platforms which give access to various courses based on different levels of certification; the quality of e-learning program and assessment parameters differ across each platform [5].

12.5.4 Online Learning Helped Us to Tackle the Crisis

In order to develop a demand supply for higher education system across the globe few measures are required to be taken. There should be improvement of quality of higher education toward global mobility of students and faculty. Moreover, there are abrupt measures to diminish the effect of pandemic on job offers, research projects, and internship programs.

The higher education sector has been significantly disrupted by the pandemic which is a critical determinant to the nation's economy [6]. A huge number of students who enrolled in abroad universities, especially in countries such as China, Australia, US, UK, etc., are worst affected by the pandemic. Many students were obstructed to leave from these countries if this situation continues eventually and then there is decrease in exhort for the international higher education is anticipated. The huge agitate on everyone is an effect of the disease on the employment rate [13]. Recent graduates are scared of withdrawal of job offers from industry due to the current situation. The crisis has reconstructed the old chalk-talk teaching to a new driven technology. This interference in conveyance of education is ensuring an inclusive e-learning solution and making a sector digitalized globally.

University is productively using tools such as Moodle, Google Meet, and Zoom to continue outcome-based teaching without any breakdown in lockdown. They directed teaching in such a way that each student and faculty has their own account in Moodle and it can be accessed anywhere using the Internet. Faculties can upload their respective PPTs, PDFs, or videos so that students can easily solve the puzzles and assignment easily. Faculties are always in touch with the parents by WhatsApp to provide all the updates. Many universities have already established swayam NPTEL chapters to conduct MOOC courses which are offered by NPTEL so that students can use this time productively. Universities have provided students and faculties to enroll various courses by Coursera. In this crisis, technology has come up with a major lifesaver. Communication plays a key role in interconnected existence and technology. Even the country is adapting a new strategy in learning; still, there is hindrance in building the endeavors completely successful. According to a survey, only 45 crore people of total population of the country can access Internet and online learning [19]. People from rural areas are still very much underprivileged of current innovations and, moreover, obstruct the importance of e-learning. Virtual classrooms require both e-learning and e-content along with online study materials, practice sheets, etc. In India, we struggle as it is not fully equipped to reach all corners of the country by digital platforms. Students in remote areas suffer and are held back due to lack of facilities, but the government is constantly trying to come up with a solution in this crisis. This pandemic is working as a tool for enhancing new methods in the education system to expand the origin and choose new techniques and platforms they have not used before. Uncertain time calls for stronger measures and our education system is stepping forward to break through the prevailing situations [19]. Using a different approach and digitalizing the sector are the two major factors which will wash away the shade of the pandemic.

A multi-dimensional strategy is required to handle the crisis and create a flexible Indian education system in long term. Instant measures are necessary to secure stability of learning universities and government schools. A learning management software and open-source digital learning solutions should be acquired so that teachers can play a key role in accessing online classes. There is rapid increase in mobile Internet users in India by approximately 90% by 2025; therefore, technology is enabling universal access and individualization of education in remote areas can be increased [18]. By this, the schooling system will be changed and there will be increase in the effectiveness in teaching methods by giving the student numerous possibilities.

12.6 Conclusion

Traditional knowledge of India is well-known across the world for Indian values, scientific innovations, and to develop feasible medicine and technologies. This study has indicated various impacts of COVID-19 on higher education in India. The pandemic established an opportunity for the introduction of virtual education and pedagogical approaches in all levels of education. In this situation, a versatile and effective educational methodology is required for sustainable development of the young minds. It will enhance their skills which build their productivity, employability, health, and well-being for the coming decades and indirectly secure overall progress of India. UGC and MHRD have launched many digital platforms with eBooks and online repositories online teaching/learning methods. Blending of traditional technologies with the digitalized technology helps students learn and access in a more comfortable way. Virtual education is most preferred during the time of the crisis. The post-COVID-19 time will completely adopt the virtual learning which also makes the students life flexible. This work has not covered any statistical analysis of COVID-19, which will be undertaken for future work.

12.7 Acknowledgment

I would like to express my deep gratitude to Dr. D. N. Reddy, Director, CR Rao Advanced Institute of Mathematics, Statistics & Computer Science for giving me this great opportunity to do this work.

References

[1] Crawford, Joseph, Kerryn Butler-Henderson, Jürgen Rudolph, Bashar Malkawi, Matt Glowatz, Rob Burton, Paulo Magni, and Sophia Lam. "COVID-19: 20 countries' higher education intra-period digital pedagogy responses." *Journal of Applied Learning & Teaching* 3, no. 1, pp. 1-20, 2020.

[2] Toquero, C. M. "Challenges and opportunities for higher education amid the COVID-19 pandemic: The Philippine context." *Pedagogical Research* 5, no. 4, 2020.

[3] Peters, Michael A., Hejia Wang, Moses Oladele Ogunniran, Yingying Huang, Benjamin Green, Jasmin Omary Chunga, Eric Atta Quainoo et al. "China's internationalized higher education during COVID-19: collective student autoethnography." *Postdigital Science and Education*, 2020.

[4] Burgess, Simon, and Hans Henrik Sievertsen. "Schools, skills, and learning: The impact of COVID-19 on education." *VoxEu. org* 1, 2020.

[5] Bao, Wei. "COVID 19 and online teaching in higher education: A case study of Peking University." *Human Behavior and Emerging Technologies* 2, no. 2, pp. 113-115, 2020.

[6] Jacob, Ogunode Niyi, I. Abigeal, and A. E. Lydia. "Impact of COVID-19 on the Higher Institutions Development in Nigeria." *Electronic Research Journal of Social Sciences and Humanities* 2, pp.126-135, 2020.

[7] Varshney, Deepanjana. "Impact of COVID-19 on Higher Education Teaching Delivery: Faculty perceptions." In *Technium Conference*, vol. 6, pp. 29-08. 2020.

[8] Sahu, Pradeep. "Closure of universities due to Coronavirus Disease 2019 (COVID-19): impact on education and mental health of students and academic staff." *Cureus* 12, no. 4 , 2020.

[9] Dwivedi, Yogesh K., D. Laurie Hughes, Crispin Coombs, Ioanna Constantiou, Yanqing Duan, John S. Edwards, Babita Gupta et al. "Impact of COVID-19 pandemic on information management research and practice: Transforming education, work and life." *International Journal of Information Management* 2020.

[10] Serafini, Gianluca, Bianca Parmigiani, Andrea Amerio, Andrea Aguglia, Leo Sher, and Mario Amore. "The psychological impact of COVID-19 on the mental health in the general population.", pp. 531-537, 2020.

[11] Ting, Daniel Shu Wei, Lawrence Carin, Victor Dzau, and Tien Y. Wong. "Digital technology and COVID-19." *Nature medicine* 26, no. 4, pp. 459-461, 2020.

[12] Gonzalez, Teresa, M. A. de la Rubia, Kajetan Piotr Hincz, M. Comas-Lopez, L. Subirats, S. Fort, and G. M. Sacha. "Influence of COVID-19 confinement in students performance in higher education." *arXiv preprint arXiv:2004.09545,* 2020.

[13] Kramer, Amit, and Karen Z. Kramer. "The potential impact of the COVID-19 pandemic on occupational status, work from home, and occupational mobility." 2020.

[14] Ahel, Oliver, and Katharina Lingenau. "Opportunities and Challenges of Digitalization to Improve Access to Education for Sustainable Development in Higher Education." In *Universities as Living Labs for Sustainable Development*, pp. 341-356. Springer, Cham, 2020.

[15] Crawford, Joseph, Kerryn Butler-Henderson, Jürgen Rudolph, Bashar Malkawi, Matt Glowatz, Rob Burton, Paulo Magni, and Sophia Lam. "COVID-19: 20 countries' higher education intra-period digital pedagogy responses." *Journal of Applied Learning & Teaching* 3, no. 1 pp. 1-20, 2020.

[16] Iivari, Netta, Sumita Sharma, and Leena Ventä-Olkkonen. "Digital transformation of everyday life–How COVID-19 pandemic transformed the basic education of the young generation and why information management research should care?." *International Journal of Information Management*, 2020.

[17] Robbins, Tim, Sarah Hudson, Pijush Ray, Sailesh Sankar, Kiran Patel, Harpal Randeva, and Theodoros N. Arvanitis. "COVID-19: A new digital dawn?." ,2020.

[18] Fagherazzi, Guy, Catherine Goetzinger, Mohammed Ally Rashid, Gloria A. Aguayo, and Laetitia Huiart. "Digital health strategies to fight COVID-19 worldwide: challenges, recommendations, and a call for papers." *Journal of Medical Internet Research* 22, no. 6, 2020.

[19] Oldekop, Johan A., Rory Horner, David Hulme, Roshan Adhikari, Bina Agarwal, Matthew Alford, Oliver Bakewell et al. "COVID-19 and the case for global development." *World Development*, 2020.

13

Virtual Reality: Solution to Reduce the Impact of COVID-19 on Global Economy

Sushma Malik[1], Anamika Rana[2]

[1]Institute of Innovation in Technology & Management, Janakpuri, Delhi 110058, India
[2]Maharaja Surajmal Institute of Technology, Janakpuri, Delhi 110058, India
Corresponding Author: Anamika Rana, anamika_rana@yahoo.com

Abstract

COVID-19 or coronavirus has changed or modified day-to-day activities. This deadly disease has influenced all the sectors of countries of the world. Cold, cough, bone pains, and problems in breathing are the common symptoms of this viral infection. The only solution to prevent this infection is social distancing and also emphasis to take precautions like regularly washing of hands with soap or using the sanitizer, using tissues during sneezing and coughing, avoiding face-to-face communication, and wearing of a mask. Virtual reality (VR) is a technology that allows the user to interact with the computer-simulated environment and it is powerful and interactive. VR creates an imaginary world by simulating the real environment by using computer technology. With the help of VR, the user can experience the scariest and toughest situation by playing safe and with a learning perspective. In VR, users are immersed and able to interact with the 3D world instead of viewing a screen. It is the way to experience and feel the past, present, and future and also the way to create our world on customized reality. VRs play an essential role to fight this pandemic with an audio–video virtual environment. VR technology designs and develops a platform to diminish the face-to-face communication between the users. COVID-19 affects all sectors like education, business, health, politics, sports, and many more. With the help of VR, all the sectors or domains can perform or implement their functions with the usage of applications like GoogleMeet, Zoom, YouTube, Portals, etc. This article highlights the main sectors overwhelmed by a coronavirus and how VR helps to reduce the extents of coronavirus.

13.1 Introduction

The novel coronavirus, COVID-19, originated from Wuhan and has stretched fast across the world. The World Health Organization (WHO) declared Coronavirus disease 2019 (COVID-19) a pandemic in the world [1]. The new infection was spread by the virus of corona family, and, at present, the world is at a threat of COVID-19. From December 2019, countries of the world have observed the huge cases of COVID-19. The person with less immunity, old age, and any medical problems, especially linked with lungs, are mostly targeted by this disease. No vaccine is available for COVID-19 till now. This virus disease is transmittable and that is why the figure of people infected with this disease is increasing rapidly day by day. The transmission of this virus is spread from person to person and also occurs in community transmission. Coronavirus can spread a number of diseases in which respiratory disorder is one of them. Indication of coronavirus may come up within two weeks after the infection. According to WHO, the most common indication of coronavirus includes fever, dry cough, and tiredness. Serious symptoms of COVID-19 which cause death may include difficulty in breathing, chest pain, loss of speech, or movement. The only prevention is the main concern to manage COVID-19. The cases of coronavirus are increasing daily. COVID-19 affects the daily life of human beings. It mainly affects the health, economic, and social life of human beings [2]. It gives the mental stress to human being. In the present scenario, all the domains are affected by COVID-19.

Due to this pandemic, globally, there are big losses which cannot be exactly estimated. The various domains move virtually to survive through virtual reality (VR) technology. VR can be defined as designing and developing the simulated environment which is, to some extent, like the real-time situation. Three dimension worlds of computer graphics have formed by ourselves who we desire without having any borders or constraints and we can improve it with our imagination which acts as the fourth dimension [3]. But as the nature of human beings, it is not enough – they want more. Instead of just watching a picture on the screen, people want to become a part of this imaginary world and want to intermingle with it. This wish of the human being is solved by technology that becomes more popular and fashionable in the current era, which is called VR [4]. COVID-19 will change the world into VR, and after this pandemic time, models of "work from home" are implemented and business travels are limited with virtual meetings. Many sectors are now moving from a physical platform to a virtual platform for survival as shown in Figure 13.1 and also increase the economy of the world. In a few months, this virus spread globally and become the pandemic globally. The spread of the virus is very fast due to the mobility of people globally by crossing international borders. The coronavirus spreads at a very high rate from human to human contact.

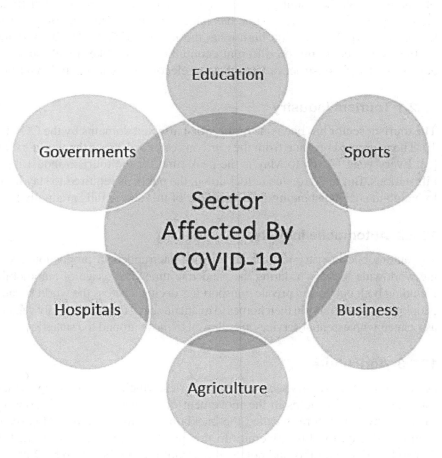

Figure 13.1 Major domains using virtual reality.

13.2 Impact of Covid-19 on the Economy

Only social distancing is identified as the best possible method to put the brakes on the spread of COVID-19 due to deficiency in treatment and medicine. Governments of different countries imposed curfew and lockdown to keep the distance between humans and tried to stop the infection of the virus. This period of lockdown marked a huge dent in the global economy. In the lockdown period, all nonessential services and all the trade sectors have been forced to shut down which causes significant disturbance in the supply chain, also putting millions of people at high risk of losing their jobs. Due to lockdown, restrictions are implemented to stop the trade of the majority of goods across the country borders, and due to these, international trades are collapsed. This epidemic time has paralyzed the global economy during this period [5]. This section highlights the impact of COVID-19 pandemic on the overall economy on different economic sectors like travel, business, education, food, construction, healthcare, telecommunication, and aviation.

13.2.1 Tourism Industry

The tourism sector has been one of the worst affected domains by the COVID-19. The generated revenue from the tourism sector is 10% of the world's GDP [5]. Every year, March to May is the peak time for the tourism industries in hill stations. But in 2020, due to lockdown, the public is required to stay home to maintain social distancing. The economy of this sector falls gradually high.

13.2.2 Automobile Industry

The automobile industry has shown a major interruption in production due to the worldwide lockdown during the pandemic time. The usage of automobiles including both public and private transport has declined across the world because people are forced to stay in their homes to maintain social distancing. Only vehicles associated with essential services are allowed to move around the states [5].

13.2.3 Agriculture

Agriculture is the primary sector of the countries, which generates employment for people and also helps in the movement of the entire circle of economy. That is why agriculture is called the backbone of any economy. This sector is more affected by this epidemic. In this pandemic time, a global crash in demand from restaurants and hotels of agricultural supplies drop by 20% [6]. Farmers are facing problems like shortage of pesticides and fertilizers because of the effect of COVID-19 on global trade. The shortage of migrant laborers has affected the functioning in various parts of the agricultural sector [7].

13.2.4 Aviation Industry

This pandemic has a massive impact on the aviation sector. Globally, all the affected countries with this situation have been forced to ban traveling of both international and national passengers. Only aircrafts with the essential services are allowed to move [5]. After this situation, many airports are now having zero or near to zero passengers traveling through these terminals. A number of countries have implemented travel restriction from one state to another state and also banned the movement across the border of the country. It also lacks clarity when these restrictions will be removed. According to International Air Transport Association (IATA), airline revenue has been dropped globally by USD 314 billion in this pandemic time [8].

13.2.5 Oil Industry

Lockdown and ban on the movement of national and international passengers across the world have resulted in a drastic decline in the consumption of fuel and a sharp decline in the global oil demand [5]. Demand for fossil fuel, except for LPG, has declined because, in India, all forms of public transport like rail, road, and air have been suspended. In that situation, the refiners are forced to cut down their throughput to up to 25%–30%. COVID-19 has an impact on the price and trade of crude oil and has also resulted in Brent crude price reaching the lowest in 17 years [9].

13.2.6 Construction Industry

In the COVID-19 pandemic time, construction industries face several troubles and delays in running projects. Due to lockdown, laborers are not able to move from one place to another. Also due to self-quarantine guidelines, construction firms will be required to stop all nonessential operations. This will result in the large-scale rescheduling of running projects, which leads to lots of losses for the industry [5]. Direct impact from the slowdown of goods on hand and labor with lockdown and in some time terminations of parties or entire projects[10].

13.2.7 Food Industry

Coronavirus effects are also seen in the food industry. Indians are foodies and they like street food more. But due to lockdown, people are not allowed to move outside and this causes the decline of revenue. In this situation, restaurants, cafes, and other luxury food services have been forced to shut down. Only grocery stores are open to fulfill the demand but are unable to meet the rising demand in masses [5]. Dining is not allowed in restaurants and cafes, which causes a major loss in revenue.

13.2.8 Healthcare and Medical Industry

Hospitals and medical industries are worst affected by this coronavirus. Due to this, it took a huge toll on healthcare professionals in developed countries like Italy, Spain, France, and the USA. Healthcare experts fear that developing countries are also facing the same problem where the result can be worst in the absence of proper health infrastructure. In that situation, normal surgeries are being delayed for patients because doctors and nurses are not able to give their hundred percent because of less liquidity and work overload. Health professionals are working 24 hours without even taking a one-minute rest to save lives and serve their duties [11].

13.3 Domain Moving Toward Virtual Reality for Survival

13.3.1 Education

All the educational institutions are temporarily shut down by the Government to prevent the spread of the coronavirus pandemic. In India too, the government has closed all educational institutions which affect the students from school level to postgraduate students and there is no declaration when they will reopen. This time is very hard for the education sector because, in this time, a number of activities like school admissions, examinations, entrance tests of various universities for their courses, and competitive examinations are held. E-learning or virtual learning process tries to solve this problem. Several online-class sites provide and offer free online classes or provide attractive discounts on e-learning modules like Nptel, Unacademy, Coursera, etc., for higher education, and Byjus, Khan Academy, Vedantu, Topper, and many more for classes up to 12. After closing of physical centers of education like schools, colleges, and even tuition centers, the usage of these sites increased. With the use of sites and applications, students can take their classes virtually without moving from one place to another. As shown in Figure 13.2, Virtual platforms are used by education centers for a number of academic meetings, seminars, and conferences through various applications like GoogleMeet, Zoom, GoToMeeting, and many more. Due to coronavirus, the educational institutes have become virtual institutes,

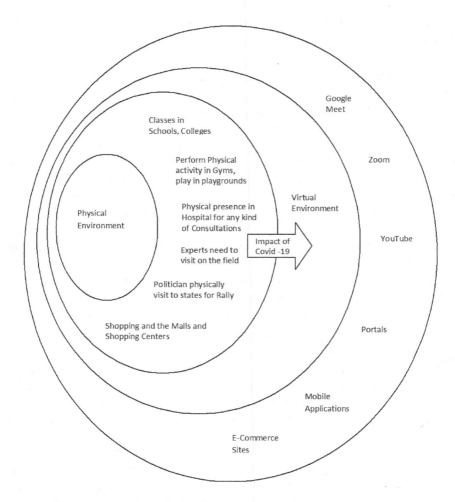

Figure 13.2 Transformation of domains from physical to virtual.

and home of every student has become his institute. Nowadays, webinar is the mostly used platform by the universities and colleges. A webinar is the grouping of two words "web" and "seminar." A webinar can be defined as an event that is held on the Internet and attended by online viewers. It is also called the online seminar and virtual event. It is the one-to-many communication method in which the presenter can contact with the bulky and specific group of online viewers from the present location. YouTube has also become the main source of education for students of schools and colleges. Students can see the number of related videos of searched topics and can do online studies. YouTube is an ocean to find new things and learn from that and is also a source of entertainment [12].

13.3.2 Hospital

The condition of the health sector is becoming tough during the COVID-19 pandemic. Due to coronavirus, a large figure of doctors and health workers lost their lives caring and treating COVID-19 patients. Doctors, administrative, or health workers are not able to give their hundred percent because of less liquidity and work overload. Health professionals are working 24 hours without even taking a one-minute rest to save lives and serve their duties. Digital Health and Telehealth come in use in the current situation. Due to COVID-19, patient's diagnoses have shifted from physical to virtual treatments [11]. Now, many health workers are giving consultation using remote methods like video conferencing and phone calls. Social distancing is the only way to stop the escalation of the coronavirus; a number of doctors are using remote methods to contact their patients. Indian Government has developed a mobile application named AarogyaSetu to fight against COVID-19. The main motive of this app is to provide the risk information and best practices against COVID-19 to the user [13].

13.3.3 Agriculture

During the pandemic period, farmers are not moving and take suggestions from the crop experts if their plants are infected with some kind of diseases. So E-plant clinics replaced the physical appearance of farmers into the virtual platform and provide effective plant healthcare to the farmers. YouTube Channels, TV programs like Krishi on Doordarshan, and E-Plant clinics are available to assist the farmers. On the portal of the virtual clinic, the farmer just needs to upload the picture of infected crops, and after analysis, experts will recommend the solution to the farmer. Experts could expand their research by using the E-clinics mode to share information [14].

13.3.4 Sports

The pandemic of COVID-19 is also affecting the sports sector. It can affect the sportsperson to media coverage. A number of sports events have been canceled or suspended to reduce the increase of virus and even the Olympics and Paralympics have been postponed and will be held in 2021 for the first time in history. During this pandemic time, to stop the spread of the virus, all places for physical activity are closed, such as fitness centers, parks, gymnasium, stadiums, and playgrounds [15]. The impact of this is less physical activity and irregular sleep which causes weight gain and defeat the physical fitness and also increases the mental stress. In this time, a number of fitness studios minimize the subscription rates to apps for smartphones and for online videos which change the physical activity from physical to virtual environment.

13.3.5 Businesses

Businesses face the major impact of this pandemic. In this time, a number of organizations do not perform their work or complete their remaining work. It becomes more challenging for businesses to continue with the same speed as earlier during this time because of less revenue. Businesses have only one solution at this time which is to become or create the virtual business, meaning transfer their business from a physical environment to a virtual environment. Many IT companies allowed their employees to work remotely, meaning work from home, during this time to end the escalation of coronavirus. Emails and online meetings or conferences with a number of mobile applications are helping to communicate and eliminate the physical communication. Call centers are other firms that go virtual, and several organizations like IBM and JetBlue permit their staff to work from home. Another example of virtual business is E-commerce sites. These sites are available from the last many years, but their usage has increased during this pandemic period. From the E-commerce sites, the user can purchase anything from a needle to gold ornaments.

13.3.6 Government

A government is the system or group of people who performs specific responsibilities and functions that should be carried out daily. The government can be defined as "By the People and For the People." To implement their functions, government bodies are performing a number of activities like rallies, meeting with higher authorities, and meeting with party members. Before this pandemic, the government held the rallies and meetings in states by physical presence, but now it has changed to virtual meetings. During the period of coronavirus, the ruling Bharatiya Janata Party (BJP) holds virtual rallies and meetings to maintain social distancing to diminish the escalation of coronavirus. Honorable Prime Minister of India conducts the South Asian Association for Regional Cooperation (SAARC) states meeting through video conference because of COVID-19. It is the first time that this kind of international meeting was held virtually [16, 17].

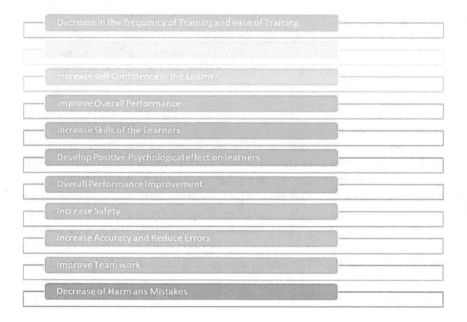

Figure 13.3   Merits of virtual reality in the battle with COVID-19 pandemic.

VR has opened a new invention in the digital era. It is rapidly changing the traditional way of executing the task into virtually learning methodologies and new approaches to handle the cases. The VR concept implements computer-based engineering and techniques to develop a virtual working environment. VR has a number of concepts and processes that help in diminishing the effect of COVID-19. Figure 13.3 shows the various merits that represent the execution of VR in the battle with the COVID-19 pandemic.

13.4 Challenges During Implementation of Virtual Reality

VR has been discovered as a new invention of the binary world. VR can be defined as a developing and designing simulated expertise which is, to some extent, similar to the real-time situation. This technology gives a number of applications in various domains, and these applications are used in daily work. But, as usual, every coin has two sides; one side of VR is very useful in every domain, but, on the other side, it faces several challenges. Figure 13.4 shows some challenges for implementation of Virtual Realityin the following [18].

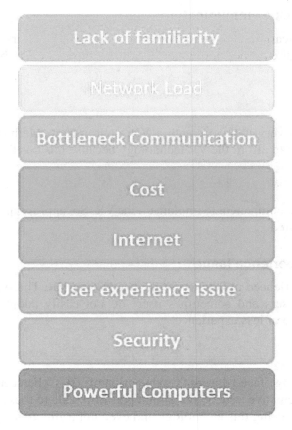

Figure 13.4 Challenges Faced by Virtual Reality implementation.

13.4.1 Lack of Familiarity

Users are less familiar with the VR applications and their usage. The value of VR is best understood when mass users adopt this technology.

13.4.2 Network Load

In the VR environment, user data is streaming live over the Internet with minimum local caching of frequently used data. It is not possible to use a network proxy service to minimize the network load when a number of people are using it at the same location as when used for group activities due to the proprietary communications protocols.

13.4.3 Bottleneck Communication

In VR, communication can be occurring in between servers and the asset cluster and can be bottlenecks which cause the problems in communications.

13.4.4 Cost

The initial cost is very high to create a VR environment. Medium-scaled people are not able to afford the VR technology because it is costlier.

13.4.5 Internet

Internet is the main component to implement the VR. So, appropriate Internet bandwidth is required.

13.4.6 User Experience Issue

Users are not experienced in how to use the handsets of VR. These handsets are new for the users and sometimes users do not easily adopt the new technology because of less experience.

13.4.7 Security

It is another challenge faced during the adaptation of VR. Cybersecurity and data privacy is a sensitive issue in the digital platform. Due to these reasons, some users are not using this technology and even not trying to adopt.

13.4.8 Powerful Computers

To use the VR technology and to create a virtual environment, the main requirement is high powerful computer systems with a high power processor with high configuration.

13.5 Road Map Toward Normal During Covid-19

The main themes that represent the individual, organizational, and social level practices to get ready for the new normal life by COVID-19 are:
- adapting information behavior;
- developing a digital environment;
- maintaining social distancing.

During the COVID-19 pandemic, everyone needs to adopt the new information environment. Everyone is required to adapt to the new information environment during this pandemic time. Online activities are performed to access the online information about the COVID-19 related, evaluate, and also sharing processes but the emergence of online technologies is creating new challenges. These new online environments facilitate individuals globally to survive in this pandemic time and also prepare for the post-pandemic and future global crises. The information access and sharing behavior of people are affected during this disaster. Worldwide, a new information environment may become normal for people because of the unpredictable duration of the pandemic and also due to the adaptation of this environment. For example, the E-learning platform will become the common platform for education in the post-COVID-19 situation because it works remotely. E-Health and work from home are also some examples of the new information environment. Many organizations are using the digital platform in their workplace before the COVID-19 pandemic, but the outbreak of this pandemic forced most organizations to adopt the digital platform. Organizations implemented a number of digital infrastructures and tools to provide uninterrupted services to their customers. During this pandemic, eight E's are implemented by the use of the digital platform. E-health, E-learning, E-commerce, E-enterprise, E-entertainment, E-environment, equality, and E-employment are implemented during the post-pandemic time. Globally, every country implements social distancing to flatten the curve of the COVID-19 pandemic, slow down the spread of the virus, and also to reduce the pressure on the healthcare systems. Till now, no vaccine of COVID-19 is available; that is why social distancing needs to be implemented for a long period due to the contagious nature of coronavirus. To minimize the spread of COVID-19, a number of measures of social distancing are implemented, like lockdown, quarantine, and shutting down of public places. But social distancing becomes the cause of mental health problems of people like anxiety, stress, and depression because of the loss of social connections and interactions [19].

13.6 Implications for Research

Companies are expected to ramp up automation with the usage of AI applications during and after the outbreak of COVID-19. This section highlights some considerations for the development and deployment of new technology in the situation of COVID-19 [20]. Some of them are listed in Table 13.1.

Table 13.1　Implications for research.

S. No.	Implication	Description
1	Future research addressing the integration of IT strategy with business strategy	The integration of IT with business strategy could evaluate crisis management successfully and also applicable for planning to handle the future pandemic.
2	Future research considering unpredictable, unexpected, and biased results of AI	Limited academic research is conducted on understanding the impact of AI applications through methods and theories. With the potential for unpredictable, unexpected, and biased results in an environment within which development and deployment are hastened by necessity, future research to increase this understanding becomes critical. AI technology helps to create new biases in judicial decision making.
3	Future research on repurposing AI to find quick solutions	AI technology plays a vital role to find quick solutions and alerts during the COVID-19. For the quick solutions during the pandemics and other crises, AI technology yields more benefits.
4	Future research on data	More research has been done on the data quality focus on the collection, solutions, and methods for the assessment of data quality. The action of any technology should depend on the quality of data. So, data should be error-free which is required for the implementation.
5	Future research on diversity in AI team membership	Based on the specific purpose of technology, future research could examine the contributions and capabilities of both the development and deployment of team members.

13.7 Conclusion

COVID-19 pandemic has covered almost the whole world and the numbers of patients are increasing day by day as of now since December 2019. The main target of this coronavirus is people with less immunity, who are aged, and having any medical problems especially related to lungs. Till now, no vaccine has been discovered for COVID-19. It spreads with physical contacts; so the only solution to minimize the infection is social distancing and usage of the virtual technology of the computer science field. This technology plays an important role in everyone's life, from school going students to the Prime Minister of India for performing their tasks by using applications or portals. COVID-19 affects the economy of the world by affecting a number of domains like tourism, automobiles, food, aviation, agriculture, health sector, and many more. But, during this time, virtual technology plays a vital role to improve the economy of each and every domain. For the implementation of education, VR technology is used to continue the study with online classes by using a number of applications and portals. In the health domain, telehealth and online consultancy are used to minimize the burden of health workers during this pandemic time. In the agriculture sector, E-clinics are used by the farmers to get expert suggestions by uploading the photos of infected plants. Sports activities are also moving from a physical environment to a virtual environment by using mobile applications. Politicians are also conducting virtual rallies to communicate with the party members and to the citizens. VR technology becomes a vital part of life during this pandemic time, but, on the other hand, VR also faces a number of challenges in the implementation, such as 5G internet, lack of familiarity, cost of implementation, and usage. Powerful computers and security are also some challenges that are faced during the implementation and usage of VR technology.

References

[1] N. Montemurro, "The emotional impact of COVID-19: From medical staff to common people," Brain. Behav. Immun., vol. 1591, no. March, pp. 1–2, 2020, doi: 10.1016/j.bbi.2020.03.032.

[2] A. Haleem, M. Javaid, R. Vaishya, and S. G. Deshmukh, "Areas of academic research with the impact of COVID-19," Am. J. Emerg. Med., 2020, doi: 10.1016/j.ajem.2020.04.022.

[3] S. Mandal, "Brief Introduction of Virtual Reality & its Challenges," Int. J. Sci. Eng. Res., vol. 4, no. 4, pp. 304–309, 2013.

[4] K. M. Menon and A. Srivastava, "Virtual campaigns - UP Front News - Issue Date: Jun 22, 2020," Jun. 2020. .

[5] V. Chamola, V. Hassija, V. Gupta, and M. Guizani, "A Comprehensive Review of the COVID-19 Pandemic and the Role of IoT, Drones, AI, Blockchain, and 5G in Managing its Impact," IEEE Access, vol. 8, no. April, pp. 90225–90265, 2020, doi: 10.1109/ACCESS.2020.2992341.

[6] M. Nicola et al., "The socio-economic implications of the coronavirus pandemic (COVID-19): A review," Int. J. Surg., vol. 78, no. March, pp. 185–193, 2020, doi: 10.1016/j.ijsu.2020.04.018.

[7] C. Staff, "[Burning Issue] COVID-19 and its Impact on Agriculture – Civilsdaily," article, 2020. .

[8] Y. Porwal, "COVID 19 impact on Airline Industry," May 2020. .

[9] S. ,Kumar Kar, "Will the oil industry survive COVID-19 effects? - Opinion by Sanjay Kumar Kar | ET EnergyWorld," Mar. 2020. .

[10] J. P. Chivilo, G. A. Fonte, and G. H. Koger, "A Look at COVID-19 Impacts on the Construction Industry | Insights | Holland & Knight," May 2020. .

[11] S. Ghosh, "COVID-19 and its Impact in Healthcare Tech Across the Globe -," Apr. 2020. .

[12] "Impact of COVID-19 on Education System in India." https://www.latestlaws.com/articles/impact-of-COVID-19-on-education-system-in-india/ (accessed Jun. 20, 2020).

[13] G. Bora, "Coronavirus update India: Now, coronavirus-affected and suspected patients can get virtual healthcare | Coronavirus news," Apr. 2020. .

[14] A. Nagaraj, "Pests in a pandemic? India's plant doctors will see you online now - Reuters," May 2020. .

[15] S. Hall, "This is how coronavirus is affecting sports | World Economic Forum," Apr. 2020. .

[16] B. Karki, "Why Did India Decide to Activate SAARC During the COVID-19 Pandemic? – The Diplomat," Mar. 2020. .

[17] "Covid-19 will alter Indian political life - The Navhind Times." https://www.navhindtimes.in/covid-19-will-alter-indian-political-life/ (accessed Jun. 20, 2020).

[18] B. I. Fox and B. G. Felkey, "Virtual reality and pharmacy: Opportunities and challenges," Hosp. Pharm., vol. 52, no. 2, pp. 160–161, 2017, doi: 10.1310/hpj5202-160.

[19] S. L. Pan and S. Zhang, "From fighting COVID-19 pandemic to tackling sustainable development goals: An opportunity for responsible information systems research," Int. J. Inf. Manage., no. June, p. 102196, 2020, doi: 10.1016/j.ijinfomgt.2020.102196.

[20] J. C. Sipior, "Considerations for development and use of AI in response to COVID-19," Int. J. Inf. Manage., no. June, p. 102170, 2020, doi: 10.1016/j.ijinfomgt.2020.102170.

Index

About the Editors

Dr. Arpit Jain is currently working as an Assistant Professor (Selection Grade) with the Department of Electrical and Electronics Engineering in University of Petroleum & Energy Studies (UPES). He has rich experience in curriculum design and has designed the curriculum for data analytics and machine learning specializations. He received the B.E. degree from SVITS, Indore, in 2007, the M.E. degree from Thapar University, Patiala, in 2009, and the Ph.D. degree from UPES, India, in 2018. He has more than 10 years of teaching and research experience. The areas of his research interests include real-time control system, fuzzy logic, machine learning, and neural networks. He has published research articles in SCI/Scopus indexed journals.

Dr. Abhinav Sharma is presently working as an Assistant Professor (Senior Scale) with the Department of Electrical & Electronics Engineering in University of Petroleum & Energy Studies (UPES). He received the B.Tech.

213

degree from H. N. B. Garhwal University, Srinagar, India, in 2009, and the M.Tech. and Ph. D. degrees from Govind Ballabh Pant University of Agriculture and Technology, Pantnagar, India, in 2011 and 2016, respectively. He has a rich teaching and diversified research experience. The areas of his research interests include signal processing and communication, smart antennas, artificial intelligence, and machine learning. He has published research articles in SCI/Scopus indexed journals and in national and international conferences.

Dr. Jianwu Wang is an Assistant Professor with the Department of Information Systems, University of Maryland, Baltimore County (UMBC). He is also an affiliated faculty at the Joint Center for Earth Systems Technology (JCET), UMBC. He received the Ph.D. degree in computer science from Institute of Computing Technology, Chinese Academy of Sciences, in 2007. His research interests include big data analytics, scientific workflow, distributed computing, service oriented computing, etc. He has published 90+ papers with more than 1400 citations (*h*-index: 20). He is/was an Associate Editor or Editorial Board Member of four international journals, co-chair of four related workshops. He is also Program Committee Member for over 30 conferences/workshops, and a Reviewer of over 15 journals or books. Since joining UMBC in 2015, he has received multiple grants as PI funded by NSF, NASA, DOE, State of Maryland, and Industry. He is also an NSF CAREER awardee. His current research interests include Big Data analytics, distributed computing, and scientific workflow with application focuses on climate and manufacturing.

Prof. Dr. Mangey Ram received the Ph.D. degree major in mathematics and minor in computer science from G. B. Pant University of Agriculture and Technology, Pantnagar, India. He has been a Faculty Member for around 12 years and has taught several core courses in pure and applied mathematics at undergraduate, postgraduate, and doctorate levels. He is currently the *Research Professor* with Graphic Era (Deemed to be University), Dehradun, India. Before joining the Graphic Era, he was a Deputy Manager (Probationary Officer) with Syndicate Bank for a short period. He is Editor-in-Chief of *International Journal of Mathematical, Engineering and Management Sciences*, and *Journal of Reliability and Statistical Studies*, Editor-in-Chief of six Book Series with *Elsevier, CRC Press-A Taylor and Frances Group, Walter De Gruyter Publisher Germany, and River Publisher*, and the Guest Editor & Member of the editorial board of various journals. He has published 225 plus research publications (journal articles/books/book chapters/conference articles) in *IEEE, Taylor & Francis, Springer, Elsevier, Emerald, World Scientific*, and many other national and international journals and conferences. Also, he has published more than 50 books (authored/edited) with international publishers like *Elsevier, Springer Nature, CRC Press-A Taylor and Francis Group, Walter De Gruyter Publisher Germany, River Publisher*, etc. His fields of research are reliability theory and applied mathematics. Dr. Ram is a Senior Member of the IEEE, Senior Life Member of Operational Research Society of India, Society for Reliability Engineering, Quality and Operations Management in India, Indian Society of Industrial and Applied Mathematics, He has been a member of the organizing committee of a number of international and national conferences, seminars, and workshops. He has been conferred with *"Young Scientist Award"* by the Uttarakhand State Council for Science and Technology, Dehradun, in 2009. He has been awarded the *"Best Faculty Award"* in 2011; "Research Excellence Award" in 2015; and, recently, *"Outstanding Researcher Award"* in 2018 for his significant contribution in academics and research at Graphic Era Deemed to be University, Dehradun, India.